'Contrarianism, the conscious denial of evidence in support of sectional interests, sets the scene for extremist authoritarian solutions to climate change. This important book demonstrates how associated far-right survivalism – based on exclusion, white supremacy and a fortress mentality – reverberates through the cybersphere and resonates at ground level among those desperate for answers to global heating. Challenging eco-fascism means knowing where it comes from, how it has evolved and how best to combat it. This book is a must read for those pursuing social and ecological justice'.

Rob White, *Emeritus Distinguished Professor of Criminology, University of Tasmania, Australia*

'The current climate crisis has seen the far right in Australia (and around the world) promote neo-Malthusian solutions wrapped up in survivalism, white supremacy and bioregionalism. *Global Heating and the Australian Far Right* shows how many on the Australian far right have long incorporated concepts of nature into their political outlook, with the Australian landscape seen as something to be both dominated and used to nourish white settler colonial society. Richards, Brinn and Jones expertly outline the various ways in which environmental and anti-environmental politics have been embraced by the far right and the threat they present in the era of dramatic climate change'.

Evan Smith, *Lecturer in History, Flinders University, Adelaide, Australia; author of* No Platform: A History of Anti-Fascism, Universities and the Limits of Free Speech *(Routledge, 2020)*

Global Heating and the Australian Far Right

Global Heating and the Australian Far Right examines the environmental politics of far-right actors and movements in Australia, exploring their broader political context and responses to climate change.

The book traces the development of far-right pseudo-environmentalism and territorial politics, from colonial genocide and Australian nationalism to extreme-right political violence. Through a critical analysis of news and social media, it reveals how denialist and resignatory attitudes towards climate change operate alongside extreme-right accelerationism, in a wider Australian political context characterised by reactionary fossil fuel politics and neoliberal New Right climate change agendas. The authors scrutinise the manipulation of environmental politics by contemporary Australian far- and extreme-right actors in cross-national online media. They also assess the political-ideological context of the contemporary far right, addressing intergovernmental approaches to security threats connected to the far right and climate change, and the emergence of radical environmentalist traditions in 'New Catastrophism' literature. The conclusion synthesises key insights, analysing the mainstreaming of ethnonationalist and authoritarian responses to global heating, and potential future trajectories of far-right movements exploiting the climate crisis. It also emphasises the necessity for radical political alternatives to counter the far right's exploitation of climate change.

This book will be of interest to researchers of climate change, the far right, neoliberal capitalism, extremism and Australian politics.

Imogen Richards is a criminology lecturer at Deakin University, Australia, where she also researches comparative forms of political violence. Her first book, *Neoliberalism and Neo-jihadism*, explored the propaganda and financial practices of neo-jihadist organisations, and her second book, *Criminologists in the Media*, examined the public scholarly practices of criminologists.

Gearóid Brinn is a PhD candidate and teaches political theory at the University of Melbourne, Australia. His research focuses on political radicalism, especially anarchism, environmentalism and realist political philosophy. His work has appeared in the *European Journal of Political Theory* and *Environmental Politics*.

Callum Jones is a researcher and PhD candidate at Monash University, Australia, whose research focuses on political extremism, particularly the networks and discursive strategies of radicalised groups and the violence they produce. His wider research focus extends to other ideological groups, including religious extremists and members of the Manosphere.

Routledge Studies in Fascism and the Far Right

Series editors: **Nigel Copsey**, *Teesside University, UK* and
Graham Macklin, *Center for Research on Extremism (C-REX),
University of Oslo, Norway*

This book series focuses upon national, transnational and global manifestations of fascist, far right and right-wing politics primarily within a historical context but also drawing on insights and approaches from other disciplinary perspectives. Its scope also includes anti-fascism, radical-right populism, extreme-right violence and terrorism, cultural manifestations of the far right, and points of convergence and exchange with the mainstream and traditional right.

Memory in Hungarian Fascism
A Cultural History
Zoltán Kékesi

The Right in the Americas
Distinct Trajectories and Hemispheric Convergences, from the Origins to the Present
Edited by Julián Castro-Rea and Esther Solano

Far-Right Ecologism
Environmental Politics and the Far Right in Hungary and Poland
Balša Lubarda

The Right and the Nation
Transnational Perspectives
Edited by Toni Morant, Julián Sanz and Ismael Saz

Metapolitics, Algorithms and Violence
New right activism and terrorism in the attention economy
Ico Maly

Plínio Salgado
A Brazilian Fascist (1895-1975)
João Fábio Bertonha

Global Heating and the Australian Far Right
Imogen Richards, Gearóid Brinn and Callum Jones

For more information about this series, please visit: www.routledge.com/Routledge
-Studies-in-Fascism-and-the-Far-Right/book-series/FFR

Global Heating and the Australian Far Right

Imogen Richards, Gearóid Brinn and Callum Jones

Routledge
Taylor & Francis Group

LONDON AND NEW YORK

First published 2024
by Routledge
4 Park Square, Milton Park, Abingdon, Oxon OX14 4RN

and by Routledge
605 Third Avenue, New York, NY 10158

Routledge is an imprint of the Taylor & Francis Group, an informa business

© 2024 Imogen Richards, Gearóid Brinn and Callum Jones

British Library Cataloguing-in-Publication Data
A catalogue record for this book is available from the British Library

Library of Congress Cataloging-in-Publication Data
A catalog record has been requested for this book

ISBN: 978-1-032-34980-0 (hbk)
ISBN: 978-1-032-35131-5 (pbk)
ISBN: 978-1-003-32543-7 (ebk)

DOI: 10.4324/9781003325437

Typeset in Times New Roman
by Deanta Global Publishing Services, Chennai, India

Contents

Acknowledgements

This research monograph only came to fruition by efforts related to the situations it describes, from environmental advocacy to the diligent work of historians and researchers documenting the historical development of far-right social movements in Australia. The book speaks to these events occurring alongside advancements in (anti-)environmental politics and the growing threats of the climate crisis. In addition to incorporating original empirical data and analysis, the forthcoming discussions draw upon significant existing information repositories on political and environmental subjects, as well as analytical insights on relevant events established in prior research.

We would like to express our gratitude to our supportive colleagues at Deakin University, Monash University and the University of Melbourne for their ongoing encouragement and collaboration. We especially appreciate the feedback on chapter drafts and encouraging words from Reece Walters, Laura Bedford, Mark Wood, Fethi Mansouri and Brian Hughes, as well as from others both within and outside academia. At this point, we acknowledge and thank the activists whose work forms the basis of much of the forthcoming critical analysis. We are particularly grateful to Andy Fleming, Cam Smith and the dedicated investigators at the White Rose Society for sharing information and insights about the Australian far right, without which this book would not have been possible.

A heartfelt thank you to our families, friends and comrades for your interest in the book and continued support throughout the writing process. We greatly appreciate the professional copy-editing work of Julia Farrell. Much of the research for this book also occurred during a postdoctoral fellowship provided by Deakin University and hosted at the Alfred Deakin Institute for Citizenship and Globalisation. The Institute has been very supportive of our research and the manuscript's development, and we thank our respective universities and the Institute for providing the resources needed to ensure the project's successful and timely completion.

We acknowledge that some chapters in this book contain revised versions of material previously published in journal articles. Chapter 3 includes reworked content from 'Ecofascism online: conceptualising far-right actors' response to climate change on Stormfront' (2022) in *Studies in Conflict and Terrorism*, and Chapter 4 features an amended excerpt from 'Newsmaking criminology in Australia and New Zealand: results from a mixed methods study of criminologists' media engagement' (2020), published in the *Australian and New Zealand Journal of*

Criminology. Additionally, Chapter 5 contains a revised fragment from 'Capturing the Environment–Security–Development nexus: intergovernmental and NGO programming during the climate crisis' (2023), published in *Conflict, Security, and Development*. We are grateful to the anonymous reviewers of this work, who directed us towards useful scholarship and whose recommendations allowed us to further develop the ideas explored in this book. Lastly, we extend our thanks to the far right and fascism series editors for their faith in the research programme set out in this book, investigating the increasingly mainstream impact of the intersections between global heating and the far right.

Acronyms and abbreviations

AAFI	Australians Against Further Immigration
AAS	Australian Academy of Science
ABC	Australian Broadcasting Corporation
AFM	Australia First Movement
ALP	Australian Labor Party
ALR	Australian League of Rights
ANA	Australian Natives' Association
ANM	Australian Nationalist Movement
ANR	Australian National Review
ASLE	Association for the Study of Literature and Environmental Leadership
AWU	Australian Workers' Union
BOM	Bureau of Meteorology
CSIRO	Commonwealth Scientific and Industrial Research Organisation
CO$_2$	Carbon dioxide
CPA	Communist Party of Australia
CRIS	Centre for Resilient and Inclusive Societies
CWR	Culture War Resource
DNC	Democratic National Committee
EAM	European Australian Movement
ENR	European New Right
ESD	Environment–Security–Development (nexus)
GAR	Global Assessment Report on Disaster Risk Reduction
GHG	Greenhouse gas
GSG	Global Scenarios Group, Stockholm Environment Institute
IPA	Institute of Public Affairs
IPCC	Intergovernmental Panel on Climate Change
ITS	Individuals Tending toward Savagery/Tending to the Wild
IWW	Industrial Workers of the World
LDC	Less developed country
LNP	Liberal party and National party coalition – 'Liberal-National Coalition'
MDG	Millennium Development Goal
NA	National anarchist

NCC	National Civic Council
NGO	Non-government organisation
NRANZ	New Right Australia/New Zealand
NRF	National Revolutionary Faction
NSN	National Socialist Network
NSP	New Sustainable Paradigm
ONP	One Nation Party
OSB	Operation Sovereign Borders
PETM	Paleocene-Eocene Thermal Maximum
PIA	Pre-industrial averages
PKK	Kurdistan Workers' Party (Partiya Karkerên Kurdistanê)
PLE	Pioneer Little Europe
RSL	Returned and Services League
UAP	United Australia Party
UPF	United Patriots Front
WMO	World Meteorological Organization
WST	World Systems Theory

1   The far right and the environment in Australia

Introduction

This book explores far-right environmentalist ideologies developing in the context of climate change. The far-reaching consequences of climate change have significantly impacted both human and non-human lives, giving rise to such extreme weather events as rising sea levels, intense heatwaves, droughts, bushfires, powerful storms and hurricanes. Among the many negative social and environmental outcomes of these events are irreversible biodiversity loss, vulnerable infrastructure, the spread of disease, and increased food and water insecurity, along with social precarity fuelling patterns of political violence on a global scale (Beeson 2010; Forchtner 2019). In response to these circumstances, far-right actors and governments have capitalised on public concern about the environment to advocate for authoritarian and ethnonationalist climate solutions. Far-right groups have also increasingly targeted populations and individuals displaced by climate change, who are often referred to ambiguously and apolitically as 'environmental migrants' or 'climate refugees' (Hartmann 2010; Gemenne 2011; Coole 2021).

Far-right environmentalist positions arise from political–ideological environments in which the media and political actors in the Global North erroneously blame the Global South for climate change through accusations of overpopulation and excess fossil fuel use (Moore and Roberts 2022). As right-wing individuals and governments increasingly endorse extreme population control measures (Biehl and Staudenmaier 2011), far- and extreme-right accelerationist actors advocate for the collapse of societies worldwide (Anker and Witoszek 1998). Exclusionary responses to climate change within Global North countries with European colonial histories – or 'the West' – have also been influenced by lifeboat ethics, which posit that wealthy states have the moral right to limit international aid, prevent the poor from accessing their resources, and oppose immigration (Hardin 1974; Neumayer 2006). This perspective remains prevalent, despite the fact that wealthy states historically significantly contributed to global heating through their reliance on resource-intensive industries (Hickel et al. 2022), while many countries in the Global South lack the levels of industrialisation necessary for sustainability in the context of global heating.

Reflecting on this situation, this book investigates the role that climate change and environmental politics more broadly play in the ideologies and propaganda

DOI: 10.4324/9781003325437-1

of Australia-based far-right actors, in both institutional and extra-institutional set-tings. The forthcoming chapters employ discourse and documentary methods to critically analyse how the far right in Australia engages with a wide range of envi-ronmental politics, in mediatised, political–economic contexts, and in light of the impacts of neoliberal capitalism.

The investigation emphasises that right-wing environmental politics in Australia are critically influenced by the country's colonial legacy, which includes the genocide of Indigenous peoples by British settlers and their dispossession of land that had been cared for and managed for over 65,000 years (Reynolds 2006; Langton 2020). Although the subject throughout this book is modern 'Australia', the analysis recognises the country's status and label as a British colonial con-struct. The borders and sovereignty imposed by colonisers continue to negatively impact Indigenous peoples and their cultures, including through processes of criminalisation and dispossession, and by drastically restricting the possibilities for Indigenous sovereignty and self-determination (Watson 2014; Moreton-Robinson 2015; Cunneen 2020). The analysis set out in this book is also based on the fact that the founding logic of Australia's colonial federation of states was apparent in the *Immigration Restriction Act 1901*, which, combined with the *Pacific Island Labourers Act 1901*, was known as the 'White Australia policy'. Indeed, a land treaty with Indigenous Australians is still non-existent, and the policies that shaped Australia's contemporary political environment are seen to be uniquely predicated on an explicit form of European white supremacy enshrined in Australia's border policies (Davis and Williams 2021).

Key concepts and global trends

With the distinct Australian experience in mind, it is important to consider key international factors influencing the Australian far right. These include the rise of the far right and extreme right globally, as well as emerging anti-knowledge and 'post-truth' politics in contemporary news and social media environments (Davis 2021). Post-truth in this context refers to the dismissal of established knowledge and evidence, typically for the purpose of reinforcing social hierarchies or promot-ing political agendas (McIntyre 2018; Biesecker 2018). Post-truth and the related concept of 'fake news' are, however, contested terms, given that both lack stable meaning, are attached to injurious politics, and problematically imply the existence of inclusive and representative 'real news' (Mondon and Winter 2020). As this book seeks to show, these concepts feature in counter-intellectual, anti-knowledge media environments where far-right ideas on the environment circulate, but struc-tural bias and political–economic imperatives continue to underwrite both legacy and new forms of media (Habgood-Coote 2019).

Far-right ethnonationalist and authoritarian responses to destructive climate impacts also function within international ideological frameworks that increas-ingly emphasise the need for a transformative reorganisation of human societies to effectively address the causes and consequences of global heating (Harvey 2005; Raworth 2017; Shiva 2018). Indeed, the severity and scale of climate change

impacts are directly related to the degree of *global heating* caused by human activities, despite the scientific consensus that global temperatures must not exceed 1.5°C
above pre-industrial levels (Mann 2021). Current emission paths are bringing the
world to a threshold of at least a 2–3°C increase by the end of the century, leading
to 'tipping points' at which irreversible damage could occur, such as melting of all
sea ice, ice caps and glaciers, resulting in long-term and significant sea-level rises,
and scenarios where plants, including food crops and even forests, would cease
to grow (Intergovernmental Panel on Climate Change (IPCC) 2022). In this context have emerged realpolitik threats of statist violence over potentially dwindling
resources (Schwartz and Randall 2003), in countries where the poorest and most
vulnerable populations, particularly in the Global South, are already suffering the
consequences of climate change and will continue to do so in the future (Bulkeley
2013).

Australia's contribution to the greenhouse gas (GHG) emissions driving climate change should not be understated. One recent study revealed this contribution, examining 'fair shares' of resource use determined by a global sustainability
corridor, based on the notion of an equally distributed ecological commons across
the global population (Hickel et al. 2022). In the study, fair shares were subtracted
from countries' overall resource use, including raw material consumption or the
'material footprint' over the course of production chains, from the period 1970–
2017. High-income countries, which represent only 16% of the global population,
were found to be responsible for 74% of resource use in excess of fair shares and
were therefore revealed as the primary drivers of global environmental degradation, representing a process that the report's authors termed 'ecological colonisation' (Hickel et al. 2022). The United States (US), progenitors of the global 'war on
terror', was responsible for 27% of the global annual overshoot of material excess
resource use, the European Union (EU) was responsible for 25%, while Australia
was the single largest of the 15 largest overshooters per capita, emitting almost 2%
or 30 tonnes per capita (Hickel et al. 2022).

Greenhouse gas emissions from human activities are a primary driver of global
heating, along with other lesser contributing factors like deforestation, land-use
changes, agricultural activities, industrial processes and waste management (IPCC
2014). Despite this fact, many political actors from both the left and the right have
incorrectly and often without justification blamed climate change on what is often
referred to as 'overpopulation'. The overpopulation theory, which posits that an
increasing global human population leads to greater demands on resources and
associated environmental impacts (Ehrlich 1968), is often applied on the political right to disproportionately target people in developing countries with predominantly non-white populations, ignoring the over-consumption and resource use in
wealthier states with significant 'white' populations (Sasser 2018). On the left, this
theory has also caused harm by overlooking the primary drivers of global heating,
which stem from industrial activity.

The previously discussed lifeboat analogy developed by Hardin (1974) emerged
from a right-wing understanding of sustainability, related to climate change and
population reduction theory. The analogy likens climate change impacts to a

situation in which individuals on a lifeboat, representing people in a safe and habitable Earth, must limit the number of people they save to prevent capsizing. While the notion informs potential contemporary policies on responses to climate change, it also has its roots in the origins of Western environmentalism. The US eugenics movement of the early twentieth century aimed to improve genetic composition through selective breeding, with influential figures such as President Theodore Roosevelt and eugenicist Madison Grant emphasising the connection between racial purity and nature (Chase 1977). Additionally, the scientific field of ecology played a role, as the concept of 'ecological succession' suggested that ecosystems developed linearly towards an 'ideal state' (Clements 1916).

Although Hardin's influence was primarily on the right, it is also important to recognise other enduring influences on environmental thought across the political spectrum. Liberal biologist Paul Ehrlich also wrote a highly influential book that shaped modern-day environmentalism: the 1968 publication *The Population Bomb*. This book primarily attributed ecological destruction to an alleged problem of overpopulation (Ehrlich 1968). Mistakenly predicting worldwide famine and upheaval in the forthcoming decades, the book was based on field research and Ehrlich's personal experience of crowding while visiting Delhi, despite the fact that Paris had a greater population density at the time of his visit (Sabin 2013). Ehrlich's theories have continued to inform contemporary references to population as a primary driver of climate change across the liberal popular media, including David Attenborough's 2020 *A Life on Our Planet* and Michael Moore's 2019 documentary *Planet of the Humans*. Overpopulation theory, when viewed through this lens, has also been used to justify dehumanising population control policies targeting poor and marginalised communities in developing states, such as forced sterilisation or coercive family planning programmes historically run by powerful countries like the US (Connelly 2008).

The origins of the 'populationist' ideologies we discuss here can also be found in the influential eighteenth-century English economist Thomas Malthus. Throughout numerous works spanning several decades, Malthus developed his theory that population growth would surpass the availability of resources, resulting in widespread poverty, famine and other societal issues. For example, in his 1798 publication, *An Essay on the Principle of Population*, Malthus posited that positive checks on population growth should encompass factors that raise mortality rates, such as war, famine and disease, which take effect when resources are scarce. According to Malthus, preventative measures, in the form of 'checks', involve voluntary actions individuals can take to reduce the number of children they have, such as postponing marriage or practising abstinence. He called for the abolishment of welfare policies and other forms of assistance for the poor, contending that these policies perpetuated dependency cycles and deterred individuals from implementing preventative measures. In essence, Malthus advocated allowing disease and deprivation to ravage underprivileged communities to safeguard middle-class living standards (Malthus 1798). In modern ecofascist interpretations of these politics, like Pentti Linkola's viewpoint, Hardin's lifeboat analogy is employed with a neo-Malthusian focus on a global scale to justify an overtly exterminationist agenda of extensive population reduction (Linkola 2009).

For the purposes of the following discussion, it is also relevant that sustainability, in the context of populationist theories on climate change, refers to a form of 'social reproduction', in the preservation of a specific social order through the continuation of economic systems, as well as the social relations and ideologies that maintain and perpetuate existing social hierarchies and inequalities (Bourdieu 1977). Acknowledging the need for change due to current detrimental conditions, theories of sustainability often concentrate on reconfiguring existing institutions and systems (Redclift 2005). Advocates of 'sustainable development', as embodied in the United Nations (UN) 2030 Sustainable Development Goals (SDGs) agenda aim to preserve global capitalism – seen here as continued economic growth and its attendant processes of production and consumption – by addressing the economic aspects of environmental conservation and emphasising the potential destruction of natural systems caused by unrestricted market access (MEA 2005). Other interpretations of *ecological* sustainability have by contrast striven to ensure the sustainability of specific ecological communities and their inhabitants, frequently drawing on the struggles, knowledge and experiences of Indigenous peoples (Gibson-Graham 2006).

Progressive approaches to sustainability increasingly emerge from post-capitalist perspectives, which argue that tackling climate change requires seeking to shield the environment from commodification. This viewpoint reflects the contemporary extension of a historical development, originating with the development of capitalist production in England from the late medieval period, and the commodification and alienation of 'the Commons', including the privatised enclosure of common lands from the sixteenth to the nineteenth centuries (Polanyi 1944).[1] In the context of this book's discussion, it is also important to note that both mainstream left- and right-wing political ideologies have addressed sustainability issues from a security-oriented standpoint, with the goal of maintaining stable, state-based social relations (Dalby 2002). Far-left ideological approaches tend to highlight the possibility of achieving ecological sustainability through social justice and systemic change (Peet et al. 2011), while far-right perspectives often concentrate on less effective, socially exclusionary and ethnonationalist responses to climate change.

Sustainability is thus a disputed concept, inherently beset by conflict and tension (Adams 2008). In this light, the forthcoming discussions in our book underscore discursive and practical engagements by right-wing actors with a core focus on 'sustainability' – specifically, a focus on the feasibility (or lack thereof) of industrial society or capitalism, and whether such systems can endure despite their social and environmental repercussions (Goodland and Daly 1996). Acknowledging the critical role of the relationship between humanity and non-human nature for sustainability, it is also vital to recognise that the 'environment' itself is indispensable for the continuation of economies (Lélé 1991). Sustainability has emerged as a form of compromise, with an ideological core centred on ongoing resource use, but with persistent debate regarding its practical implications (Lélé and Norgaard 2005). In this context we can consider the implications of similar, related buzzwords; for instance, numerous individuals labelled as 'conservatives' in Australia and elsewhere are, in reality, statist reactionaries (Davis 2019).

As examined in this book, far- and extreme-right actors, such as ecofascists, often view sustainability not solely (or necessarily) as the preservation of eco-logical balance and resource conservation but also as the safeguarding of assumed racial or cultural purity (Biehl and Staudenmaier 1995). This mindset was exempli-fied in the infamous '14 words' slogan of David Lane, the leader of the former mur-derous US white supremacist group, The Order: 'We must secure the existence of our people and a future for white children'. Besides emphasising 'overpopulation', ecofascists and green authoritarians frequently make false claims that immigration and cultural mixing – or 'racial miscegenation' – are the main drivers of environ-mental degradation and resource depletion (Dyett and Thomas 2019). Using these justifications, far-right political–ecological actors often advocate for strict border controls, population control measures and even ethnic cleansing to achieve their warped vision of sustainability (Forchtner 2019). As this book aims to reveal, these actors do not arise in an ideological vacuum but instead reflect the political and historical–ideological contexts that fuel populationist and sustainability narratives across the political spectrum (Staudenmaier 2021).

Considering the significant overlap between Global North supremacy and the advance of global heating, we adopt a multifaceted approach when exploring the intersection of Australian far-right actors with environmental politics, including climate change. The analysis first recognises how many far-right actors exploit the issue of climate-related and other forms of migration, portraying immigrants and refugees as threats to national resources and security. Additionally, many of these actors deny or downplay the reality of climate change, aligning their actions and interests with extractive industries that profit from environmental destruction. This book also acknowledges that the anti-democratic and authori-tarian tendencies of far-right actors hinder meaningful climate change action, as they seek to undermine effective mitigation policies that require collective action and international cooperation. The current research, therefore, seeks not only to describe the situation but also to respond to Lubarda and Forchtner's (2022) call for 'countercommunication' against the far-right's racist and nationalist environ-mental views. It investigates and offers alternative climate change response narra-tives, while also situating far-right environmentalist messages within the broader context of white supremacist ideologies to which they are affiliated (Dyett and Thomas 2019).

Another primary objective of this book is to illuminate the political field in which different climate change responses operate. The chapters thus explore cli-mate change discourses from right-wing activists and institutions, mainstream political and news media sources in Australia, and institutional and intergovern-mental networks of response to the perceived intersectional threats posed by global heating, in conjunction with the rise of a far-right social movement. By examining various typologies of climate change politics, the analysis also aims to refute the false polarities sometimes attributed to distinct radical climate response traditions (Dessler and Parson 2019). In this regard, the critical evaluation of grassroots, advocacy and activist discourses on climate change linked to the progressive left provided in the book's concluding parts unpacks climate change crisis responses that may be susceptible to far-right co-optation, as well as others that might offer meaningful alternatives to far-right environmentalist positions.

Through its comparative research methods, this project sought to examine the relationships between various institutional and non-institutional actors in the Australian environmental political landscape. While not all actors and cases studied may be strictly considered 'far-right', the investigation aimed to identify and analyse the dialectical relationships between different right-wing conservative, reactionary and revolutionary movements responding to the political pressures and societal tensions associated with climate change impacts. The study reveals instances of reciprocity, interdependence, tension and divergence in the relationships between a range of different right-wing actors. In its entirety, we hope that the investigation critically exposes the connections between disparate actors, which, within their given context, can be said to influence and represent the right's social responses to climate change on a broader scale.

The actors and movements explored in this book are often described as 'right-wing', encompassing a variety of neoliberal and Australian nativist elements. Mudde's typology (2019) is used here to classify some actors as 'far-right', including those on the radical to the more extreme ends of the right-wing spectrum. While the radical right primarily operates within liberal democratic frameworks, it also endorses anti-immigration sentiments, cultural bias towards 'Western civilisation' and the valorisation of radical traditionalism (Minkenberg 2017; see also McFadden 2023). Extreme actors are by contrast more associated with neo-fascist, accelerationist or neo-Nazi ideologies, combined with a glorification of revolutionary violence (Carter 2013). The far right sits at the interstice of the two, intersecting with and spanning elements of both. In this analysis, far-right actors are seen to engage in institutional–majoritarian mainstream politics, while also drawing on various currents of revolutionary, ethnonationalist or eugenicist political reasoning (Mudde 2019).

Given that another core objective of this investigation was to examine the growing prevalence of right-wing responses to climate change in contemporary society, we also adopted a definition set out by Mondon and Winter of the far right as comprising 'movements and parties that espouse a racist ideology, but [who] do so in an indirect, coded and often covert manner, notably by focusing on culture and/or occupying the space between illiberal and liberal racisms, between the extreme and the mainstream' (2020, 21). In line with this definition, the research as a whole aims to scrutinise how actors based in Australia leverage environmental and pseudo-environmentalist discourses to bolster their white supremacist, far-right belief systems, potentially facilitating the greater acceptance of these beliefs in mainstream society.

Climate change and the far right in Australia

The intersection of global heating and the far right holds particular significance for Australia, given its devastating experience of climate change, restrictive immigration policies and far-right political violence emanating from the country's political situation. This environment was evidenced in one high-profile case involving the actions of Brenton Tarrant, an Australian self-described ecofascist, who perpetrated a mass shooting at two mosques in Christchurch, New Zealand, in 2019, leading to 51 deaths (Macklin 2019). While Tarrant was inspired by rhetoric about a 'global population cap' in the so-called manifesto of Anders Breivik (2011), the neo-Nazi who murdered 77 in Norway in 2011, his actions also went on to inspire other violent white supremacists (Bangstad 2014). Among others, these included Patrick Crusius, who perpetrated a 2019 mass shooting in Texas against Hispanic and Latino people that killed 23, naming his manifesto after Al Gore's climate change documentary (Smith 2021). Another recent illustration was the alleged actions of Payton Gendron, who targeted black people in a mass shooting at a Buffalo supermarket in 2022 that led to the death of 10, and who also described his attack as in the interests of 'green nationalism' (Abbas et al. 2022; Anon 2022).

Despite the recent attention given to the ecological–environmental concerns expressed by white supremacist actors, their claims are often relatively uncritically accepted in academic, journalistic and policy forums (Thomas and Gosink 2021). Owing to a lack of understanding of ecofascism and far-right 'ecologisms' generally, right-wing governments' anti-immigration policies in response to climate impacts are also sometimes described vaguely as fascistic (Sassen 2018). Empirical research is also lacking on the impact of far-right political entryism in domains of environmental advocacy, and the spread of ecofascism and other far-right ecological ideas through government policies and popular discourse (Forchtner 2019). These areas for research are globally relevant, but are also especially pertinent in Australia, in light of its future experience of expected extreme weather impacts and the prevailing political environment in the country, where ethnocultural exclusion and supremacy, particularly in respect of asylum seekers and immigration, remain prevalent (Hage 2017).

Australia's vulnerability to climate change therefore forms a critical background to this investigation. The country has routinely experienced frequent and severe natural disasters, such as heatwaves, droughts, bushfires and floods, resulting in loss of human life, mass death of non-human species, infrastructure damage and economic catastrophes (Flannery 2008), and these have been intensifying in recent years, as exemplified in the 2019–2020 'Black Summer' bushfires. Australia is also internationally associated with the immense risk posed to one of the world's famous natural wonders, Queensland's Great Barrier Reef, owing to rising ocean temperatures and coral bleaching caused by global heating (Hughes et al. 2015). While Australia has long been known for its historical reliance on coal for energy production, making it one of the world's highest per-capita GHG emitters (Hickel et al. 2022), Australian political leaders have faced criticism for their slow progress in reducing emissions and ongoing support for the fossil fuel industry (Lawrence and Laybourn-Langton 2022). At the time of writing, the federal government's emission reduction targets are widely recognised as incommensurate with international efforts to limit global heating to below 2°C, as defined in the Paris Agreement (Ivanovski and Churchill 2020; IPCC 2023).

In light of these dynamics, the book surveys climate change accelerationism by self-proclaimed eco- and neo-fascists in Australia, but also analyses resignatory and denialist attitudes towards climate change expressed by a wider spectrum of right-wing Australian political actors. With a critical lens, the account of news and social media presented in this book emphasises the far-right reaction to environmental advocacy in Australia, and its influence on government policy. This influence is viewed again in the context of the country's controversial response to climate change, which has been marked by New Right neoliberal incursions into political decision-making, and powerful politicians and media outlets downplaying or denying the severity of climate change (Hamilton 2007; see Chapter 4 p. 135).

Despite the growing exploitation of structural, historical and material inequalities arising from climate change by far-right elements in Australia and beyond, there has been a noticeable absence of focus on far-right activities outside the realms of 'immigration' or 'cultural politics' (Lubarda and Forchtner 2022). In the context of Australia, there is a similar lack of understanding and attention to the ways in which environmental politics can be manipulated by globally oriented far-right groups. These actors also take advantage of developmentalism and climate change in other European settler nations (Richards 2023).

Much of the existing research on white supremacist far-right activism in Australia has examined its targeting of Indigenous people, individuals with Asian and African cultural backgrounds, people from Jewish and Muslim communities, and people belonging to other non-dominant religious and cultural groups (Akbarzadeh 2016; Peucker and Smith 2019). There has also been more limited research on the intersecting discrimination and violence enacted by far-right actors against individuals from social strata marginalised on grounds other than religion or 'race', including women, non-heterosexuals, and people with disabilities (Dunn et al. 2004). This focus has been understandable, given the dominant localised expressions of the Australian far right, and their white supremacist orientation. However, as in relation to some other states with European histories, the relatively exclusive focus of the research on far-right targets and methods of organisation provides limited understanding of the broad social impact of organised far-right political activity, and the ways in which its cultural expressions find purchase among susceptible recruits.

Green nationalism, fossil fascism and ecofascism

Focusing on the relationships between different far-right expressions of environmental politics in Australia, this book proceeds with an awareness that different traditions of right-wing climate response internationally have been conceptualised in a number of ways. Within and beyond governments, environmentalism on the far right has been examined as a part of wider radical nativist agendas, focusing on the politics of individual countries or within specific political parties often defined as 'populist' (Copelovitch and Pevehouse 2019). A key focus in this literature has been the recurrent tension between conservative climate-sceptic political groups, conspiracy theories, and the acceptance of climate science among some nationalistic political parties and their support-ers, with homesteading and agriculture also being embraced in some far-right civic groups (Hamilton 2002). One factor seen to prevent the solid adoption of prevention and mitigation measures by governments has been the prevalence of anti-'global elitism' in populist right narratives (Lockwood 2018), combined with conspiratorial fears of climate science as a 'socialist plot' (Hatakka and Välimäki 2019).

These trends, though, are not homogeneous or consistent. Several scholars emphasise how European far-right parties in particular are inconsistent on the issue of climate change and environmentalism, with alliances between green and nationalist parties often being short-lived and situationally contingent (Fowler and Jones 2005; Fowler and Jones 2006; Karbevski 2020). While the rise of the right in Europe has generally been thought to drive climate-sceptic and anti-ecological positions (Jeffries 2017), it also included European far-right reaction to refugees entering the continent in the wake of the 2015 Syrian Civil War, acts of political violence by neo-jihadist organisations and their affiliates, and the rise of authori-tarian regimes in the region. These events are collectively sometimes seen to drive a backlash against 'multiculturalism', while also fuelling parochial resistance to collective or meaningful climate action (Meyer-Ohlendorf and Görlach 2016).

While the centre, traditional or populist right of the right-wing spectrum are usually interpreted as more likely to be climate sceptics (Hatakka and Välimäki 2019), those on the further right, bearing 'blood and soil' perspectives that refer to a racially defined national body connected to a specific geographic area, are more likely to accept the reality of climate change (Forchtner et al. 2018). Rising accept-ance of climate change and the role of environmental thought in specific far- and extreme-right nationalistic narratives has also been recently and popularly defined as 'green nationalism' (Conversi and Friis Hau 2021). Even within the emergent environmentalist frameworks of some extreme-right parties across Europe, how-ever, these parties are still more likely than left-wing environmentalists to focus on conservation, nature preservation for human use, national character, clean living, and back-to-the-land romanticism, rather than the issue of climate change mitiga-tion. Similarly, in the US, the extra-parliamentary alt-right has adapted superficial ecothemes that accept the reality of climate change as part of a phenomenon Rueda (2020) terms 'neoecofascism'. While their concerns with respect to the environ-ment tend to be focused on preservation for human ends, rather than predicated on any intrinsic concern for 'nature', these actors also assert a kind of traditional fas-cist naturalism, placing the 'eco' under 'humans' in the natural hierarchy (Rueda 2020).

Andreas Malm and The Zetkin Collective's book *White Skin, Black Fuel* out-lines another conceptual framework for understanding the relationships between different far- and extreme-right actors responding to climate change, which spe-cifically focuses on how the consequences of one response can give rise to other reactionary strands. They introduce the concept of 'fossil fascism' to describe how the fossil fuel industry and its political allies hinder climate action, deny climate science, and attack environmental regulations to maintain their power and interests.

Fossil fascism creates conditions that enable the emergence of 'green national-ism' and 'ecofascism' due to various interconnected factors, such as the dependence of fossil fascism on authoritarian regimes that support exploitative and environ-mentally destructive capital accumulation and lead to environmental degradation and increasingly severe climatic events, which in turn provoke reactions from green nationalists and ecofascists. Ecofascism in the context of climate change can then involve both palin*defensive* ultranationalism and palingenetic ultranation-alism, according to Malm's (2021) adaptation of Griffin (2018). Palindefensive ultranationalism in this case signifies the defence of nativist, racially pure identi-ties, symbolised through the preservation of local natural environments, while pal-ingenetic ultranationalism complementarily refers to the long-held fascist ideology of an imagined ideological and physical rebirth of states along racial–national lines (van den Heuvel 2020)

While green nationalism has not been as prominent in Australia's institutional politics compared to other countries, it can perhaps be observed through the grow-ing acceptance of climate change science and a sense of parochial resignation expressed in some political circles (see Chapter 4, p. 134). Certain acceleration-ist tendencies on the extreme right might align with a form of ecofascism rooted in Western civilisation and adapted to the Australian national context (McFadden 2023), while the longstanding denialism in Australia has historically supported what Malm might refer to as fossil fascism. However, as Chapter 4 explores, the authoritarian governance structures typically associated with fossil fascism have been relatively absent in the Australian context.

Governments in Australia have been able to continue to support fossil fuel industries, including through the annual provision of million-dollar subsidies and carbon-trading schemes, due to the successful strategic public relations campaigns of conservative reactionaries and New Right lobby groups, convincing Australians that their prosperity depends on unfettered mining, especially of fossil fuels (Brett 2020). It is nevertheless important to acknowledge that with the recent introduction of repressive legislation and policing powers suppressing civil protest against the large-scale extraction, use and exploration of fossil fuels, Australia might now be witnessing a shift towards more authoritarian governance, leaning towards 'fossil fascism'. This political development might be said to constitute a 'creeping' effect, as defined by Ross (2017), where fascist ideologies gradually infiltrate and influ-ence mainstream politics and culture.

Dominant responses to climate change within Australian political and media institutions are then relatively distinctive from other international examples in large part because they are influenced by the country's introduction of climate change mitigation measures designed to ensure the long-term viability of fossil fuel industries. As mentioned, these measures have included carbon storage and offset schemes, as well as large government fossil fuel subsidies, many of which have been beset by corruption controversy and widely criticised for their ques-tionable efficacy in reducing GHG emissions (Shahbaz et al. 2017). The preser-vation of fossil fuel trading in Australia has been rhetorically and ideologically justified by the country's history of extractive colonialism, and practically justified by proponents due to the nation's economic reliance on the mining and export of natural resources (Banerjee 2000).

While Malm's (2021) framework therefore has the potential to illuminate relationships between the various intersecting responses of right-wing and far-right actors to climate change, including in Australia, it also emphasises the fundamental role of the fossil fuel industry and its political allies in hindering meaningful climate action (see Wainwright and Mann 2018). It is also important to note that this circumstance has been replicated in various ways and to different extents globally, with denialism increasingly commonly described as giving way to different forms of ecofascism (Moore 2021). The ties between 'fossil fascism' and 'green nationalism' implicating both Australian and international actors are moreover not always nationally bound, or overt. Colonial capitalist projects predicated on the carbon economy are influenced by similar rationales, where environmental destruction incurred in one place can be 'offset' by measures such as reforestation and GHG mitigation in another place (Moore 2015). This occurs despite the fact that these measures may often be enacted in parts of the world where industrial activity generating climate change has not historically been prevalent, and despite these measures having significant deleterious impacts on the lives and sustainable livelihoods of local, often Indigenous, populations (Gilbertson et al. 2009; Kanngieser 2016).

The above activities are sometimes referred to as 'ecofascism', but in the context of the present research it is necessary to note the dominant historical usage of this term, while accepting that its meaning is still highly subjective and can have different interpretations depending on who uses it. Ecofascism is generally taken to refer to a strain of far-right ecologism predicated on eugenicism and genocidal blood and soil logics (Lubarda 2020). It is in some more lateral definitions seen to entail the enaction of 'cultural genocide' through the destruction of environments – including via climate change and development projects – necessary to sustain human societies (Thomas and Gosink 2021). In some cases, ecofascism is also taken to signify neo-Malthusian policies that facilitate atrocities and provide legitimacy to the far right, or drastic measures to securitise borders in response to climate change, constituting what Thomas and Gosink call a 'weaponisation of geographies' (2021).

In the interest of understanding how far-right environmental politics are gaining greater purchase, it is important to emphasise how the term ecofascism may be used to refer to ideological tendencies among environmentalists who oversimplify the problem to 'humans' and their population as the primary eco-issue (Hendrixson and Hartmann 2019). The nature of this intersectional prejudice becomes apparent, where misanthropic belief systems are underpinned by overpopulation narratives with racist and gendered components, blaming either humanity as a whole for the crises of the Anthropocene[2] or directly targeting 'women of the Global South' – who disproportionately experience the most devastating climate change impacts – for their reproductive capabilities (Dyett and Thomas 2019, 215).

In this context, it is necessary to acknowledge criticism of the Anthropocene concept for potentially perpetuating populationist narratives (Moore 2015). While the Anthropocene is not yet officially recognised as a distinct historical epoch (Zalasiewicz et al. 2010), the term has been widely used to convey the idea that human activities, especially since the late-eighteenth and nineteenth century Industrial Revolution, have had a profound and enduring impact on the planet, warranting the definition of a new epoch in Earth's geological history (Crutzen and Stoermer 2021). This marks the end of the Holocene, which spanned the past 12,000 years following the last Ice Age and was characterised by a relatively stable climate that facilitated advancements in agriculture, technology, civilisation and population growth, and biodiversity and ecosystem stability (Steffen et al. 2011). Despite the criticisms of the concept, it is essential to note that theories about when the 'Anthropocene' began vary, with proposed starting points including the emergence of agriculture thousands of years ago (Ruddiman 2003); the Industrial Revolution from the late-eighteenth century (Crutzen and Stoermer 2021); and the 'Great Acceleration' in the mid-twentieth century, characterised by rapid industrialisation, population growth, and GHG emissions contributing to climate change (Steffen et al. 2015).

An alternative perspective, rooted in the neo-Marxist World Systems Theory (WST) as proposed by Moore (2015), introduces the term Capitalocene. This concept shifts the focus from 'humanity' and highlights the role of the capitalist system of production and consumption in causing environmental damage, resource depletion and pollution. Like other WST approaches (Wallerstein 1974), Moore sees capitalism as a world system or global economy based on commodity exchange and unequal trade relations originating in the mid-fifteenth century. Moore believes that the Anthropocene theory, which emphasises the Industrial Revolution, is flawed because it neglects capitalism's fundamental role in shaping human–nature interactions (Moore 2015). Critics of this view, however, argue that the Anthropocene theory does not aim to provide a comprehensive account of modern human–nature relationships, but instead refers to a distinct period marked by unparalleled atmospheric carbon dioxide (CO_2) levels, artificial nitrogen, species extinction, ocean acidification, sea-level rise, ozone layer depletion, and other potentially disastrous Earth System[3] disruptions occurring today (Steffen et al. 2015).

Some scholars support the term Capitalocene but do not contest, as Moore perhaps does, the emphasis of Earth System scientists on the emergence of a new geological era characterised by the impact of unprecedented human-caused atmospheric CO_2 levels (Angus 2016). Malm (2016), for example, contends that the Capitalocene is a more fitting term to account for the differentiated responsibility of major emitters in the Global North, as part of the 'fossil economy', for driving climate change, compared to the Anthropocene. Expanding these debates and applying them to related inquiries would be valuable, but space constraints in this book preclude their further exploration here. In the following chapters, the term Anthropocene is used most often, though depending on the line of critique drawn upon. This term in our usage refers to the scientific category of increased atmospheric change as 'global heating', intensified by the activities of, first and foremost, the owners of the means of production in fossil fuel use and extraction, and large-scale industrial activities, as well as by the excessive consumption of electricity and other commodities by individuals living in affluent countries (Malm 2016).

Given the divergent interpretations of ecofascism, there has been a related divide in the extant literature on ecofascism between a focus on environmentalists who advocate or accept fascistic positions in service of a green motivation and far-right actors who use green arguments to support their anti-immigrant views (Dyett and Thomas 2019). An interpretive focus on 'green fascists' over 'fascist greens' has become more dominant in recent years, partly given the growing notoriety of far-right and white supremacist attackers with stated environmental motivations, including through the mass killings at Christchurch in 2019, at El Paso in 2020 and at Buffalo in 2022 (Ross and Bevensee 2020). The concept of ecofascism, as it informs the critical discussion within this book, can refer to ideological tendencies within groups or individuals whose belief system has a substantial environmental element, but which is otherwise nativist, xenophobic or overtly genocidal. It can also refer to fascists and other far-right actors who have disingenuous or opportunistic green elements as a central part of their ideology.

We develop this understanding while recognising how climate-sceptic anti-environmentalists may use 'fascist' terminology to refer to all environmentalists, and some ecoradicals also use similar terms to refer to other ecoradicals. For example, deep ecologists, or radical environmentalists who hold a biocentric view, are sometimes accused of having ecofascist tendencies by social ecologists (Bookchin 1987; Biehl and Staudenmaier 1995), though many deep ecologists, drawing on Næss (1973), disagree and seek to defend their biocentric approaches from accusations of ecofascism (Carter 1995; Brown 2005). Some researchers have conversely asserted that liberal green capitalist technocrats, with their focus on family planning programmes, are closer to ecofascism than are deep ecologists (Dyett and Thomas 2019). The recent shock of the global COVID-19 pandemic has also led to a rise in simplistic 'humans are the virus'– type rhetoric, with slogans like 'Save Trees Not Refugees' going viral on social media, blurring the line between fascist-first and green-first ideological typologies (Allison 2020). In popular debates, the term ecofascist is usually understood as an insult and is generally not used as a self-descriptor, except by politically violent actors; nor is it a category definition for the environmental far right used by government agencies, intergovernmental institutions or non-government organisations (NGOs).

Before proceeding with the forthcoming investigations, it is important to clarify that while most evidence-based research has demonstrated that environmentalism is not intrinsically fascist, not all far-right and fascist actors adopt cynical perspectives regarding the environment. Some, such as 'green nationalists', have taken reasoned if inadequate and bigoted positions on environmental issues. However, the reality of this situation has been somewhat distorted by its treatment in a sub-set of literature that conflates fascism and radical environmentalism, suggesting that environmentalism is proto-fascist or communistic (e.g. Zimmerman 2004).

Also relevant in this context is the ongoing criminalisation of eco-activism, which has taken several forms over time (Tayler et al. 2019). One internationally influential example was the 'green scare' of the early twenty-first century, when US Government and law enforcement agencies targeted environmental activists and groups, often accusing them of ecoterrorism or other crimes (Potter 2011; Salter

2011; Pedroni 2017; Parr 2018). Extending this history, in the wake of growing mass civil movements such as Extinction Rebellion, there has been an increased criminalisation of mainstream environmental protests, civil disobedience, and even lobbying by mainstream environmental groups (Tayler et al. 2019).[4] Aside from the formal criminalisation of eco-activism, violence against environmentalists has also been steadily increasing worldwide, with hundreds of people being murdered each year worldwide, especially in developing countries, and by right-wing regimes, particularly in South America (Butt et al. 2019).

For the purpose of this book's investigation, it is necessary to briefly observe how arguments against criminalising some eco-activists as 'terrorists', particularly in North America, have recently been running up against their limits, given the increasing incidence of eco-sabotage that has threatened or taken innocent lives. In 2020, there were more than 40 attempts at railway sabotage across Washington state, including one successful derailment that started a fire, spilled 29,000 gallons of oil, and resulted in the evacuation of 120 people, as part of a pipeline resistance effort in Canada (Beaumont 2021). A few eco-activists have also employed ter- ror tactics by targeting individuals for assassination or even expressing reckless abandon as to the murder of random innocents, as in the case of the eco-extremist group that emerged in Mexico in 2011, ITS (Individuals Tending toward Savagery/ Tending to the Wild) (Ross and Bevensee 2020; Spadaro 2020).

Although some left-leaning environmental groups are beginning to consider more extreme actions that are potentially harmful to living beings, the majority are still overwhelmingly non-violent. This remains the case even among those groups that see themselves as radical or extreme and who do not see that action will be effective without radical systemic change (Hirsch-Hoefler and Mudde 2014). Even left-wing 'anti-civilisational' anarchists, who wish to see the complete collapse of modern technological society, for example, do not endorse violence to achieve this (e.g. Zerzan 2002). The only groups that do wish for extreme change, are willing to use violence to get there and whose strategy is based on immediate 'blaze of glory' actions by individuals – rather than, for example, patiently building a mass move- ment where the potential to actually move against the system is far less immediate, as with most progressive radical eco groups – exist within the online extreme-right ecofascist milieu (Loadenthal 2022).

While left-wing environmental activism often requires personal sacrifices in consumption and long-term strategies with limited success, the extreme right relies on offering disaffected individuals a violent path to fame. Their strategy is based on providing immediate and seemingly effective direct action opportunities that feed the ecofascist narrative of 'one less human, the better' (e.g. Devi 1991; Linkola 2009). These lone acts of 'terror' usually have negligible ecological impact, while they contribute to societal destabilisation and far-right recruitment (Lamoureux 2020).

Despite the growing focus on far-right movements and solitary politically violent actors, government and intergovernmental agencies have not formally recognised far-right or fascist environmentalism as a threat. This is in contrast to the historical response to non-violent eco-activist actions such as ecotage and picketing, which were swiftly labelled as terrorist acts by governments (Vanderheiden 2008), and to the rising prevalence of right-wing political violence, which is increasingly targeted by security measures (Pantucci and Ong 2021). Specific institutional responses to the stated environmental motivations of far-right violent actors remain scarce.

Governments, including security departments like the Pentagon, have sponsored research on the security implications of climate change (Hartmann 2010). However, such studies often align with the potential correlates of far-right narratives, emphasising how the climate crisis may lead to an influx of impoverished individuals from marginalised regions, thus encouraging closed-border policies (Georgi 2019). While environmentalism is increasingly employed as a pretext for anti-immigration rhetoric, some NGOs and development agencies have advocated for seemingly humane population planning methods that are actually dehumanising, such as enforced 'family planning' in the Global South, echoing earlier methods of non-consensual sterilisation (Dyett and Thomas 2019). Through the upcoming analysis, we suggest that these approaches are also susceptible to far-right co-optation, particularly to the extent that they coincide with a surge in global refugee flows and the growing international adoption of closed-border 'fortress' strategies.

The forthcoming analysis examines the news and social media industries within their political–economic context and in light of the shifting global informational environment, which is sometimes characterised as part of a 'post-truth' era (Lubarda and Forchtner 2022). Considering far-right (pseudo-)environmentalist attitudes, the investigation explores how, in contemporary counterfactual media environments, 'truth' is often deemed significant only in terms of its commodity status and function for neoliberal systems (Biesecker 2018).

While acknowledging the unique capacity of new media to circulate sensationalist and ephemeral information, the analysis also aims to counterbalance ahistorical interpretations of the anti-knowledge political tendencies within the far right. Contrary to the belief that counterfactual informational environments are exclusively a product of contemporary politics disseminated via new media, far-right and proto-fascist movements have long engaged in various modes and methods of anti-intellectual struggle against both progressivist socialistic institutions and the positivistic influences of the Enlightenment period (Eatwell 1996). Furthermore, proto-fascist and far-right political expression is grounded in the underlying rationality of superficial survival-of-the-fittest ideologies, which are characterised by the utilitarian employment of propaganda and pseudo-science (Griffin 2013; see Chapter 6, p. 206).

The subsequent chapters employ a combination of qualitative and quantitative methodologies to examine how far-right actors based in Australia exploit global historical-material disparities in reaction to contemporary crises (Mudde 2017). Case study, discourse and documentary methods are employed selectively throughout the analysis, with the objective of evaluating the degree to which public representations of right-wing reactions to climate change in Australia contribute to their mainstream political impact (Moffitt 2016; Perry and Scrivens 2018). Additionally, the analysis considers how the viewpoints of actors labelled as 'far-right' permeate broader societies, particularly in relation to their non-violent and less explicitly political activities (Busher 2016). In pursuing this investigation, we rely on a variety of primary and secondary sources that exemplify the conspicuous demonstrations of right-wing institutional and extra-institutional actors responding to climate change, including explicitly white supremacist actors. By doing so, the study builds upon existing comparative political research that elucidates how patterns of intersectional prejudice have developed through public communication mechanisms (Lentin and Titley 2011; Deland et al. 2014).

Chapter breakdown

One limitation of this book is its lack of attention to the intersections of far-right environmentalism and climate change occurring in the Global South. The investigation instead focuses primarily on Australia-based actors and the frameworks of knowledge they access, which tend to have transnational resonance on the far right in other Global North, European settler-colonial settings (Slater 2018). This focus is relevant to shedding light, however, on the dominant possibilities for future responses to devastating climate impacts, given the disproportionate responsibility of wealthy actors in the Global North for global heating (Hartmann 2010).

In response to the environmentally focused actions of Australian far-right extremists, this investigation recognises that the primary motivation underlying most extra-institutional far-right activism is the preservation and re-establishment of 'white power', rather than a comprehensive commitment to ecological conservation. This commitment requires acknowledging the global nature of ecological issues and the interconnectedness of human and non-human entities (Ross and Bevensee 2020). The subsequent discussions in this study reflect our view that, contrary to the perspectives found in some emerging literature on far-right ecological–environmental politics, the ideological commitment of extreme right-wing actors to political violence can only be considered 'environmentally oriented' if their belief system identifies the non-white Other as a 'foreign species' or 'pollutant' that poses a racial–national threat to the 'white race' (Forchtner 2019).

With this normative positioning in mind, this introductory chapter has framed the approach of the book in the context of several primary trends in research on the far right and the environment. Drawing on a growing body of research, it has elaborated the current socio-political and environmental imperatives for researching far-right ideology and climate change in concert, including within Australia. Also discussed were several examples of existing research investigating discrete right-wing environmental political phenomena, and comparative and relational forms of analysis, which this book positions as key to understanding the 'mainstreaming' function of popular, and in particular reactionary, forms of far-right politics in contemporary crisis settings. The chapter has also highlighted the importance of understanding these connections given the elements of tension and contiguity between a spectrum of right-wing actors based in Australia who respond to climate change, and in light of the country's distinct reliance on fossil fuel industries and heightened vulnerability to future climate threats.

Chapter 2 is titled 'A heated Australian landscape: histories of environmental politics on the far right'. This chapter presents a potted history of radical, nativist and extreme-right organisations and individuals in Australia since the lead-up to Federation in 1901 who have exploited elements of environmental politics for white supremacist and white nationalist political ends. The chapter foregrounds the colonial project, entailing the systemic genocide of Indigenous peoples who had occupied the mainland of Australia for at least 65,000 years prior to European invasion, as germane in understanding subsequent historical expressions of white supremacist environmentalism (Langton 2020). Since the arrival of the First Fleet in 1788, the founding narrative of Australia was predicated on the razing of local landscapes in the service of European productive agriculture, as well as the profit-driven mining of lands and exploitation of working peoples (Rose 2004). Against this backdrop, the examples of far-right activism and ideology drawn upon in the chapter include the exclusivist protection of white workers' rights in the early labour union movement and the rejection of internationalism and socialism by imperial industrialists (Burgmann 1984).

Chapter 2 also explores how far-right environmentalist influences in Australia during the twentieth century included the white supremacist, antisemitic Australia First Movement (AFM), which joined with leaders of the Jindyworobak literary club in the interwar period in a concerted cultural campaign to develop a unique Australian nativist entity, including by way of appropriating Indigenous imagery and themes (Bird 2014). Other cases include the longest-running historical Australian far-right organisation, the Australian League of Rights (ALR), which made incursions into early organic farming communities after World War II (WWII) (Moore 2005). Anti-urban rural agitations combined with a militaristic patriotism are then discussed as key elements of Jack van Tongeren's Australian Nationalist Movement (ANM) in the 1980s (Fox 2023), while radical nationalist tendencies are shown to be associated also with the leadership and governance structures of various Australian far-right movements and groups since white settlement.

Chapter 2 also addresses National Anarchist (NA) tendencies in Australia, which to date have been marginal but also internationally resonant, and they may be more important in future. The NA groupuscules are identified as seeking substate, autonomous modes of social organisation, albeit often within masculinist, racist and authoritarian ideological frameworks (Macklin 2005). Environmental ideologies developing since the 1960s predicated on the notion that overpopulation undermines environmental sustainability are also discussed in the context of this history as important for the development of misanthropic and anti-immigrant sentiments in the Australian context, as they have been internationally (Coole 2021).

Chapter 3, titled 'Ecofascism online: Australian far-right actors' use of environmental politics on cross-national media', offers insights derived from a mixed-method analysis of social media and website platforms utilised by far- and extreme-right actors based in or focusing on Australia. The investigation encompasses relatively open, mainstream platforms such as Facebook and Telegram, as well as more closed and discrete far-right white supremacist platforms like Stormfront and Gab. Drawing upon a review of grey and scholarly conceptual literature on ecofascism (Forchtner 2019; Ross and Bevensee 2020), this chapter reflects on current understanding of contemporary extreme-right actors' key ecological–environmental priorities, particularly in relation to online propaganda. By examining Australian cases, the discussion highlights several predominant themes, including online actors' endorsement of eugenicist or Malthusian population control measures (Linkola 2009), the portrayal of regular or irregular migrants as a 'foreign species' or 'pollutant', and the expression of denialism, resignation or accelerationism in response to the most significant impacts of ecological and environmental devastation (Moore 2016).

The discussion in Chapter 3 addresses the unique role of online media in facilitating the spread of white supremacist perspectives on climate change and the natural environment given their architectural ability to facilitate the curation of debate and consensus building around controversial social justice issues (Daniels 2009). The platforms in question are known to monetise psychologically addictive material that is likely to 'go viral' by virtue of its extremist content (Conway 2020). The analysis foregrounds, then, how more generalist far-right media can provide a periphery or 'radial' influence (Eatwell 1996), appealing to susceptible recruits before directing them towards more closed platforms accessed by organised neo-Nazis and neo-fascists. The generalist far-right audiences in this investigation are shown to promote climate change denialism on white supremacist forums such as Gab and Stormfront, while organised groups across Telegram and Facebook tend to engage in different forms of environmental politics, more clearly embracing the adoption of blood and soil ideologies. One of the main groups examined is the Australian National Socialist Network (NSN), whose leaders are currently subject to various criminal prosecutions for acts of interpersonal violence and preparatory terrorism-related acts, while they have also made several high-profile attempts at recruitment (McKenzie and Tozer 2021).

Chapter 4, 'Newsmaking on the environment: climate change resignation and denial in the Australian media', explores recent dominant trends in Australian media reportage on climate change and environmental politics among both mainstream news sources and more marginal far-right alternative news (alt-news) or 'alt-right' news online outlets (Frischlich et al. 2020). The analysis takes a case study approach to understanding the evolution and ideological–political history of climate change denial and resignatory attitudes within Australian news and political arenas. The cases examined include the 2019–2020 Black Summer bushfires, where several dozen people directly died, more than 3,000 homes were destroyed and more than 20 million hectares of the Australian landscape burned (Cowled et al. 2022; Davey 2023). The event also resulted in the deaths of more than 3 billion animals (AIDR 2023), and an estimated additional 450 human fatalities resulting from long-term exposure to fire and smoke (Binskin et al. 2020; Humphries 2022). The discussion of this event considers how a pattern of climate change denialism in News Corporation (News Corp) media specifically was established through coordinated strategic messaging on the part of government and media, including through the inaccurate attribution of blame for the fires to arson (Mocatta 2021).

The analysis in Chapter 4 highlights the incidence of cross-fertilisation between Australian mainstream media and global social networks, emphasising the use of automated posting and viral hashtags by far-right actors to spread misinformation about the 2019–2020 fires (Knaus 2020). A discussion section is also dedicated to the recent historical circumstances that led to the dominance of denialist climate change politics in Australia, including the political influence of the New Right, which since the 1980s has run concerted, internationally funded campaigns to promote continued political support of the Australian fossil fuel industry, resulting in a lack of meaningful action on GHG emission mitigation (Moore 1995; Kelly 2019). The chapter lastly explores New Right and mainstream media connections with reporting on environmental issues in more marginal alt-news websites, which range from pornographic, conspiracist or Christian, to national socialist in political orientation. Among several examples, those highlighted include the *Australian National Review*, *Caldron Pool*, *Epoch Times Australia*, *Eternity News*, *Rebel News*, *The Good Sauce*, *The Unshackled* and *XYZ*.

Chapter 5, titled 'New Catastrophism and the Environment–Security– Development nexus: programming and advocacy during the climate crisis', investigates the underlying motivations behind international NGO and intergovernmental initiatives aimed at preventing political violence linked to environmental issues and developmentalism. The first section critically evaluates how Environment– Security–Development (ESD) nexus policy frameworks can contribute to the securitisation of Global South communities and generate inhumane and criminogenic effects, shaping future political violence patterns in both Global North and Global South contexts (Barnett and Adger 2007; Dalby 2013). Additionally, the discussion explores the predominance of programmes targeting neo-jihadist exploitation of environmental catastrophes, as opposed to far-right uses of environmentalism (Richards 2023). The analysis also examines how developmental strategies may inadvertently perpetuate the global conditions of material inequality and deprivation that contribute to asymmetric violence. These approaches frequently do not focus on addressing the root causes of global heating or the ecological security concerns that increase the likelihood of climate change being exploited by the far right (McDonald 2018).

The latter portion of Chapter 5 scrutinises the increasing prominence of extra-institutional and grassroots progressivist perspectives on climate change, manifested through the multifaceted phenomenon of New Catastrophism (Swyngedouw 2011; Klein 2014). This approach signifies a resurgence in environmentalist literature that emphasises the imminent danger of catastrophic climate change and advocates for radical social, political and economic restructuring of current societies to effectively confront this challenge (Bendell 2018; Oreskes and Conway 2014). The analysis highlights the tendencies in the responses of these left-wing progressivist actors that may be vulnerable to far-right co-optation, as well as those that provide meaningful alternatives to far-right environmental responses.

The final chapter of the book, Chapter 6, titled 'Far-right environmentalism in the post-truth era: global networks and future directions', adopts a comparative analysis of the key findings pertaining to Australian right-wing tendencies in climate change politics within a global context (Mudde 2017). It also investigates the generative effects of both institutional and non-institutional responses to these political trends. The ensuing discussion encourages a comprehensive outlook on various climate change reactions to illuminate discourses that are frequently interconnected and dialectically influenced (Forchtner and Kølvraa 2015). Acceptance and resignation in relation to climate change within public political discourse can be portrayed as rational when juxtaposed with climate change denialism, despite the fact that diverse actors' anticipated responses to future extreme weather events may still embody authoritarian or ethnonationalist qualities (Swyngedouw 2011). Correspondingly, as the forthcoming deliberations aim to illustrate, the repercussions of one mode of right-wing response to significant climatic occurrences often pave the way for reactions and responses from advocates of another far-right ideological tradition, which could also be socially regressive or eschatologically accelerationist in nature.

Chapter 6 demonstrates how varying currents of response arise from a political–ideological climate shaped by the dominance of 'post-truth', counterfactual media, far-right anti-intellectual inclinations, and the neoliberal commodification of knowledge (Fischer 2020). In conjunction, these trends are evidenced through the book's examination of climate policies alongside news, social media and political–institutional media, which contribute to the 'mainstreaming' of far-right environmentalist stances, including scenarios where ethnonationalism and authoritarianism are likely to gain prominence (Lubarda and Forchtner 2022). These ideologies foster acts of extreme white supremacist political violence and are manifested in climate change policies such as the growing adoption of closed-border, 'fortress policies', increasingly fallaciously justified by lifeboat ethics and ahistorical, factually inaccurate populationist arguments (Carr 2016).

Furthermore, the chapter contemplates potential avenues for future research that could elucidate the continuing trajectory of far-right social attitudes regarding climate change and political frameworks that offer genuine alternatives to far-right modes of political organisation, particularly in the post-capitalist transformation of human societies (Wright 2010). Moving forward, Chapter 6 scrutinises how future potential developments in far-right ecological thought correspond to current advancements in and connections across radical environmental theory and movements. Alongside the mainstreaming of the far right, there has also been a mainstream-radical convergence on the left, with post-capitalist responses to climate change gaining traction (Fraser 2019). This chapter proceeds to examine post-capitalist perspectives on ecological sustainability that emerge from diverse traditions of green anarchism, encompassing a broad range of viewpoints, from Bookchin to anti-civilisational anarchist, green-syndicalist, and more. The analysis presented here suggests that some of these tendencies provide meaningful alternatives to the far right, while the susceptibility to far-right co-optation of others is difficult to circumvent, considering the far-right's fundamental tactic of syncretically appropriating and adapting ideas from the left (Griffin 2013).

The final chapter also encapsulates a key aspect of the book by emphasising a fundamental paradox inherent in fascist and far-right approaches to environmental politics. This paradox is produced by far-right exclusivist and fallacious natural hierarchies that both perpetuate and are sustained by capitalist environmental degradation. As history has shown through collective movements opposing fascism and the far right, the core of an antifascist project in this context lies in exposing the contradictory nature of ecological fascism, along with its political–economic derivations.

Notes

1 Commons were communal lands and resources that were managed collectively by local communities for mutual benefit. In England, from the sixteenth to the nineteenth centuries, the process of enclosure involved the consolidation and privatisation of these common lands, leading to the rise of private property and the displacement of small-scale farmers and peasants.
2 The Anthropocene is a term used to describe a proposed new geological epoch characterised by the significant impact of human activities on Earth's ecosystems and geological processes. The term itself is derived from the Greek words d*anthropos*, meaning human, and *kasd* or recent.
3 The Earth System is made up of five major interactive parts or subsystems, including the hydrosphere, atmosphere, cryosphere, geosphere and biosphere.
4 The FBI, for example, lists Greenpeace as an eco-terrorist group, and a security briefing by the UK Home Office named Extinction Rebellion and Greenpeace as extremist organisations.

References

Abbas T, Somoano IB, Cook J, Frens I, Klein GR and McNeil-Willson R (2022) *The Buffalo attack: an analysis of the manifesto*, The International Center for Counter-Terrorism, accessed 15 January 2023. https://www.icct.nl/publication/buffalo-attack-analysis -manifesto

Adams B (2008) *Green development: environment and sustainability in a developing world*, Routledge, London.

Akbarzadeh S (2016) 'The Muslim question in Australia: Islamophobia and Muslim alienation', *Journal of Muslim Minority Affairs*, 36(3): 323–333.

Allison M (2020) '"So long, and thanks for all the fish!": urban dolphins as ecofascist fake news during COVID-19', *Journal of Environmental Media*, 1(1) Supplement 1: 4–1.

Angus I (2016) *Facing the Anthropocene: fossil capitalism and the crisis of the earth system*, New York University Press, New York.

Anker P and Witoszek N (1998) 'The dream of the biocentric community and the structure of utopias', *Worldviews: Global Religions, Culture, and Ecology*, 2(3): 239–256.

Anon (2022) *You wait for a signal while your people wait for you* [Due to the sensitivity of this content we do not include a link to the original document].

Australian Institute for Disaster Resilience (AIDR) (2023) *New South Wales, July 2019– March 2020 Bushfires – Black Summer*, Australian Disaster Resilience Knowledge Hub, accessed 1 January 2023. https://knowledge.aidr.org.au/resources/black-summer -bushfires-nsw-2019-20/

Banerjee SB (2000) 'Whose land is it anyway? National interest, Indigenous stakeholders, and colonial discourses: the case of the Jabiluka uranium mine', *Organization & Environment*, 13(1): 3–38.

Bangstad S (2014) *Anders Breivik and the rise of Islamophobia*, Bloomsbury Publishing, London.

Barnett J and Adger WN (2007) 'Climate change, human security and violent conflict', *Political Geography*, 26(6): 639–655.

Beaumont H (29 July 2021) 'The activists sabotaging railways in solidarity with Indigenous people', *The Guardian*, accessed 1 June 2022. https://www.theguardian.com/environment /2021/jul/29/activists-sabotaging-railways-indigenous-people

Beeson M (2010) 'The coming of environmental authoritarianism', *Environmental Politics*, 192: 276–294.

Bendell J (2018) *Deep adaptation: a map for navigating climate tragedy*, University of Cumbria, accessed 1 January 2023. https://insight.cumbria.ac.uk/id/eprint/4166/1/ Bendell_DeepAdaptation.pdf

Biehl J and Staudenmaier P (1995) *Ecofascism: lessons from the German experience*, AK Press, Edinburgh.

Biehl J and Staudenmaier P (2011) *Ecofascism revisited: lessons from the German experience*, New Compass Press, Porsgrunn.

Biesecker BA (2018) 'Guest editor's introduction: toward an archaeogenealogy of post-truth', *Philosophy & Rhetoric*, 51(4): 329–341.

Binskin M, Bennett A and Macintosh A (2020) *Royal commission into natural disaster arrangements: report*, Commonwealth of Australia. https://naturaldisaster.royalcommission .gov.au/system/files/2020-11/Royal%20Commission%20into%20National%20Natural %20Disaster%20Arrangements%20-%20Report%20%20%5Baccessible%5D.pdf

Bird D (2014) *Nazi dreamtime: Australian enthusiasts for Hitler's Germany*, Anthem Press, London.

Bookchin M (1987) 'Social ecology versus deep ecology: a challenge for the ecology movement', *Green Perspectives: newsletter of the Green Program Project*, accessed 26 November 2022. http://dwardmac.pitzer.edu/Anarchist_Archives/bookchin/ socecovdeepeco.html

Bourdieu P (1977) *Outline of a theory of practice*, Cambridge University Press, Cambridge.

Brett J (2020a) 'Coal addiction comes at huge cost', La Trobe University, https://www .latrobe.edu.au/news/articles/2020/opinion/coal-addiction-comes-at-huge-cost

Brievik A (2011) *2083: a European declaration of independence* [Due to the sensitivity of this content we do not include a link to the original document].

Brown CS (2005) 'Ecofascism and the animal heritage of moral experience', *Dialogue and Universalism*, 15(7/8): 35–48.

Bulkeley H (2013) *Cities and climate change*, Routledge, London.

Burgmann V (1984) 'Racism, socialism, and the labour movement, 1887/1917', *Labour History*, 47: 39–54.

Busher J (2016) *The making of anti-Muslim protest: grassroots activism in the English Defence League*, Routledge, London.

Butt N, Lambrick F, Menton M and Renwick A (2019) 'The supply chain of violence', *Nature Sustainability*, 2(8): 742–747.

Carr M (2016) *Fortress Europe: dispatches from a gated continent*, The New Press, New York.

Carter A (1995) 'Deep ecology or social ecology?', *The Heythrop Journal*, 36(3): 328–350.

Carter E (2013) *The extreme right in Western Europe: success or failure?*, Manchester University Press, Manchester.

Chase A (1977) *The legacy of Malthus: the social costs of the new scientific racism*, Alfred A. Knopf, New York.

Clements FE (1916) *Plant succession: an analysis of the development of vegetation*, Carnegie Institution of Washington, Washington.

Connelly M (2008) *Fatal misconception: the struggle to control world population*, Harvard University Press, Cambridge.

Conversi D and Friis Hau M (2021) 'Green nationalism: climate action and environmentalism in left nationalist parties', *Environmental Politics*, 30(7): 1089–1110.

Conway M (2020) 'Routing the extreme right: challenges for social media platforms', *The RUSI Journal*, 165(1): 108–113.

Coole D (2021) 'The toxification of population discourse: a genealogical study', *The Journal of Development Studies*, 57(9): 1454–1469.

Copelovitch M and Pevehouse JC (2019) 'International organizations in a new era of populist nationalism', *The Review of International Organizations*, 14: 169–186.

Cowled BD, Bannister-Tyrrell M, Doyle M, Clutterbuck H, Cave J, Hillman A, Plain K, Pfeiffer C, Laurence M and Ward MP (2022) 'The Australian 2019/2020 black summer bushfires: analysis of the pathology, treatment strategies and decision making about burnt livestock', *Frontiers in Veterinary Science*, 9: 83.

Crutzen PJ and Stoermer EF (2021) *The 'Anthropocene'*, Springer International Publishing, New York.

Cunneen C (2020) *Conflict, politics and crime: Aboriginal communities and the police*, Routledge, London.

Dalby S (2002) *Environmental security*, University of Minnesota Press, Minneapolis.

Dalby S (2013) 'The geopolitics of climate change', *Political Geography*, 37: 38–47.

Daniels J (2009) *Cyber racism: white supremacy online and the new attack on civil rights*, Rowman & Littlefield Publishers, Washington.

Davey M (2 January 2023) 'More than 2,400 lives will be lost to bushfires in Australia over a decade, experts predict', *The Guardian*, accessed 1 April 2023. https://www.theguardian.com/australia-news/2023/jan/02/more-than-2400-lives-will-be-lost-to-bushfires-in-australia-over-a-decade-experts-predict

Davis A (2021) *Reckless opportunists: elites at the end of the establishment*, Manchester University Press, Manchester.

Davis M (2019) 'Transnationalising the anti-public sphere: Australian anti-publics and reactionary online media', in Peucker M and Smith D (eds) *The far-right in contemporary Australia*, Springer, New York.

Davis M and Williams G (2021) *Everything you need to know about the Uluru Statement from the Heart*, NewSouth Publishing, Randwick.

Deland M, Minkenberg M and Mays C (eds) (2014) *In the tracks of Breivik: far right networks in Northern and Eastern Europe*, LIT Verlag Münster, Münster.

Dessler AE and Parson EA (2019) *The science and politics of global climate change: a guide to the debate*, Cambridge University Press, Cambridge.

Devi S (1991) *Impeachment of man*, Noontide Press, Costa Mesa.

Dunn KM, Forrest J, Burnley I and McDonald A (2004) 'Constructing racism in Australia', *Australian Journal of Social Issues*, 39(4): 409–430.

Dyett J and Thomas C (2019) 'Overpopulation discourse: patriarchy, racism, and the specter of ecofascism', *Perspectives on Global Development and Technology*, 18(1–2): 205–224.

Eatwell R (1996) 'On defining the 'fascist minimum': the centrality of ideology', *Journal of Political Ideologies*, 1(3): 303–319.

Ehrlich PR (1968) *The population bomb*, Ballantine Books, New York.

Fischer F (2020) 'Post-truth politics and climate denial: further reflections', *Critical Policy Studies*, 14(1): 124–130.

Flannery T (2008) *The weather makers: the history and future impact of climate change*, Text Publishing, Melbourne.

Forchtner B (ed) (2019) *The far right and the environment: politics, discourse and communication*, Routledge, London.

Forchtner B and Kølvraa C (2015) 'The nature of nationalism: populist radical right parties on countryside and climate', *Nature and Culture*, 10(2): 199–224.

Forchtner B, Kroneder A and Wetzel D (2018) 'Being skeptical? Exploring far-right climate-change communication in Germany', *Environmental Communication*, 12(5): 589–604.

Fowler C and Jones R (2005) 'Environmentalism and nationalism in the UK', *Environmental Politics*, 14(4): 541–545.

Fowler C and Jones R (2006) 'Can environmentalism and nationalism be reconciled? The Plaid Cymru/Green Party alliance 1991–95', *Regional and Federal Studies*, 16(3): 315–331.

Fox VJ (2023) 'Fascism and anti-fascism in Perth in the 1980s', in Smith E, Persian J and Fox VJ (eds), *Histories of fascism and anti-fascism in Australia*, Routledge, London.

Fraser N (2019) *The old is dying and the new cannot be born: from progressive neoliberalism to Trump and beyond*, Verso Books, Brooklyn.

Frischlich L, Klapproth J and Brinkschulte F (2020) 'Between mainstream and alternative: co-orientation in right-wing populist alternative news media', in van Dujin M, Preuss M, Spaiser V, Takes F and Verberne S (eds), *Multidisciplinary international symposium on disinformation in open online media*, Springer, New York.

Gemenne F (2011) 'Why the numbers don't add up: a review of estimates and predictions of people displaced by environmental changes', *Global Environmental Change*, 21: S41–S49.

Georgi F (2019) 'Toward fortress capitalism: the restrictive transformation of migration and border regimes as a reaction to the capitalist multicrisis', *Canadian Review of Sociology/Revue*, 56: 556–579.

Gibson-Graham JK (2006) *A postcapitalist politics*, University of Minnesota Press, Minneapolis.

Gilbertson T, Reyes O and Lohmann L (2009) *Carbon trading: how it works and why it fails*, Uppsala: Dag Hammarskjöld Foundation, Pietermaritzburg.

Goodland R and Daly H (1996) 'Environmental sustainability: universal and non-negotiable', *Ecological Applications*, 6(4): 1002–1017.

Griffin R (2013) *The nature of fascism*, Routledge, London.

Griffin R (2018) *Fascism*, John Wiley & Sons, New York.

Habgood-Coote J (2019) 'Stop talking about fake news!', *Inquiry*, 62(9–10): 1033–1065.

Hage G (2017) *Is racism an environmental threat?*, Polity Press, Cambridge.

Hamilton C (2007) *Scorcher: the dirty politics of climate change*, Black Inc, Melbourne.

Hamilton P (2002) 'The greening of nationalism: nationalising nature in Europe', *Environmental Politics*, 11(2): 27–48.

Hardin G (1974) 'Commentary: living on a lifeboat', *BioScience*, 24(10): 561–568.

Hartmann B (2010) 'Rethinking climate refugees and climate conflict: rhetoric, reality and the politics of policy discourse', *Journal of International Development*, 22(2): 233–246.

Harvey D (2005) *Spaces of neoliberalization: towards a theory of uneven geographical development*, Franz Steiner Verlag, Stuttgart.

Hatakka N and Välimäki M (2019) 'The allure of exploding bats: the Finns Party's populist environmental communication and the media', in Forchtner B (ed) *The far right and the environment: politics, discourse and communication*, Routledge, London.

Hendrixson A and Hartmann B (2019) 'Threats and burdens: challenging scarcity-driven narratives of "overpopulation"', *Geoforum*, 101: 250–259.

Hickel J, O'Neill DW, Fanning AL and Zoomkawala H (2022) 'National responsibility for ecological breakdown: a fair-shares assessment of resource use, 1970–2017', *The Lancet Planetary Health*, 6(4): e342–e349.

Hirsch-Hoefler S and Mudde C (2014) '"Ecoterrorism": terrorist threat or political ploy?' *Studies in Conflict & Terrorism*, 37(7): 586–603.

Hughes TP, Day JC and Brodie J (2015) 'Securing the future of the Great Barrier Reef', *Nature Climate Change*, 5(6): 508–511.

Humphries A (2022) 'Australia's Black Summer bushfires were catastrophic enough: now scientists say they caused a 'deep, long-lived' hole in the ozone layer', *ABC News*, accessed 8 December 2022. https://www.abc.net.au/news/2022-08-26/black-summer -bushfires-caused-ozone-hole/101376644

Intergovernmental Panel on Climate Change (IPCC) (2014) *Climate change 2014*: *synthesis report* (eds Pachauri RK and Meyer LA), contribution of Working Groups I, II and III to the Fifth Assessment Report of the Intergovernmental Panel on Climate Change.

Intergovernmental Panel on Climate Change (IPCC) (2022) *Climate change 2022*: *impacts, adaptation and vulnerability*, accessed 19 March 2023. https://www.ipcc.ch/report/ar6/wg2/

Intergovernmental Panel on Climate Change (IPCC) (2023) *AR6 synthesis report: climate change 2023*, accessed 19 March 2023. https://www.ipcc.ch/report/sixth-assessment -report-cycle/

Ivanovski K and Churchill SA (2020) 'Convergence and determinants of greenhouse gas emissions in Australia: a regional analysis', *Energy Economics*, 92. http://doi.org/10 .1016/j.eneco.2020.104971

Jeffries E (2017) 'Nationalist advance', *Nature: Climate Change*, 7: 469–471.

Kanngieser A (2016) *Experimental politics and the making of worlds*, Routledge, London.

Karbevski B (2020) *Europe's far right and its attitude towards climate change and environmentalism* [master's thesis], Universitat Autònoma de Barcelona, accessed 3 March 2023. https://ddd.uab.cat/pub/trerecpro/2020/233051/TFM_bkarbevski.pdf

Kelly D (2019) *Political troglodytes and economic lunatics: the hard right in Australia*, Black Inc, Melbourne.

Klein N (2014) *This changes everything: capitalism vs. the climate*, Simon & Schuster, New York.

Knaus C (8 January 2020) 'Bots and trolls spread false arson claims in Australian fires 'disinformation campaign'', *The Guardian*, accessed 8 January 2023. https://www .theguardian.com/australia-news/2020/jan/08/twitter-bots-trolls-australian-bushfires -social-media-disinformation-campaign-false-claims

Lamoureux M (25 September 2020) 'Neo-Nazis are using eco-fascism to recruit young people', *Vice*, accessed 1 January 2023. https://www.vice.com/en/article/wxqmey/neo -nazis-eco-fascism-climate-change-recruit-young-people

Langton M (2020) 'Welcome to country: knowledge', *Agora*, 55(1): 3–10.

Lawrence M and Laybourn-Langton L (2022) *Planet on fire: a manifesto for the age of environmental breakdown*, Verso Books, Brooklyn.

Lélé SM (1991) 'Sustainable development: a critical review', *World Development*, 19(6): 607–621.

Lélé S and Norgaard RB (2005) 'Practicing interdisciplinarity', *BioScience* 55(11): 967–975.

Lentin A and Titley G (2011) *The crises of multiculturalism: racism in a neoliberal age*, Zed Books Ltd, London.

Linkola P (2009) *Can life prevail? A radical approach to the environmental crisis*, Arktos, Budapest.

Loadenthal M (2022) 'Feral fascists and deep green guerrillas: infrastructural attack and accelerationist terror', *Critical Studies on Terrorism*, 15(1): 169–208.

Lockwood M (2018) 'Right-wing populism and the climate change agenda: exploring the linkages', *Environmental Politics*, 27(4): 712–732.

Lubarda B (2020) 'Beyond ecofascism? Far-right ecologism (FRE) as a framework for future inquiries', *Environmental Values*, 29(6): 713–732.

Lubarda B and Forchtner B (2022) 'The far right and the environment: past-present-future', in Bruno VA (ed) *Populism and far right: trends in Europe*, EduCATT, Milan.

Macklin G (2005) 'Co-opting the counter culture: troy Southgate and the National Revolutionary Faction', *Patterns of Prejudice*, 39(3): 301–326.

Macklin G (2019) 'The Christchurch attacks: livestream terror in the viral video age', *CtC Sentinel*, 12(6): 18–29.

Malm A (2016) *Fossil capital: the rise of steam power and the roots of global warming*, Verso Books, Brooklyn.

Malm A (2021) *White skin, black fuel: on the danger of fossil fascism*, Verso Books, Brooklyn.

Malthus TR (1798) *An essay on the principle of population*, J. Johnson, London.

Mann ME (2021) *The new climate war: the fight to take back our planet*, PublicAffairs, New York.

McDonald M (2018) 'Climate change and security: towards ecological security?', *International Theory*, 10(2): 153–180.

McFadden A (2023) 'Wardens of civilisation: the political ecology of Australian far-right civilisationism', *Antipode*, 55(2): 548–573.

McIntyre L (2018) *Post-truth*, MIT Press, Cambridge.

McKenzie N and Tozer J (16 August 2021) 'Inside racism HQ: how home-grown neo-Nazis are plotting a white revolution', *The Age*, accessed 1 January 2022. https://www.theage.com.au/national/inside-racism-hq-how-home-grown-neo-nazis-are-plotting-a-white-revolution-20210812-p58i3x.html

Meyer-Ohlendorf N and Görlach B (2016) *The EU in turbulence*: w*hat are the implications for EU climate and energy policy?*, Ecologic Institute, accessed 19 March 2023. https://www.ecologic.eu/sites/default/files/publication/2016/2280-eu-crisis-background-paper_1.pdf

Millennium Ecosystem Assessment (MEA) (2005) *Ecosystems and human well-being: synthesis*, Island Press, Washington.

Minkenberg M (2017) *The radical right in Eastern Europe: democracy under siege?*, Springer, New York.

Mocatta G (18 October 2021) 'What's behind News Corp's new spin on climate change?', *The Conversation*, accessed 1 June 2022. https://theconversation.com/whats-behind-news-corps-new-spin-on-climate-change-169733

Moffitt B (2016) *The global rise of populism: performance, political style, and representation*, Stanford University Press, Redwood City.

Mondon A and Winter A (2020) *Reactionary democracy: how racism and the populist far right became mainstream*, Verso Books, Brooklyn.

Moore A (1995) *The right road?: a history of right-wing politics in Australia*, Oxford University Press, Oxford.

Moore A (2005) 'Writing about the extreme right in Australia', *Labour History*, 89: 1–15.

Moore J (2015) *Capitalism in the web of life: ecology and the accumulation of capital*, Verso Books, Brooklyn.

Moore JW (2021) *Comrades in arms with the web of life*: a *conversation with Jason W. Moore*, Humanities Commons, accessed 19 March 2023. http://doi.org/10.17613/sj6r-sm11

Moore JW (ed) (2016) *Anthropocene or Capitalocene?*: *nature, history, and the crisis of capitalism*, Pm Press, New York.

Moore S and Roberts A (2022) *The rise of ecofascism: climate change and the far right*, Polity Press, Cambridge.

Moreton-Robinson A (2015) *The white possessive: property, power, and Indigenous sovereignty*, University of Minnesota Press, Minneapolis.

Mudde C (2017) *The populist radical right: a reader*, Routledge, London.

Mudde C (2019) *The far right today*, Polity Press, Cambridge.

Næss A (1973) 'The shallow and the deep, long-range ecology movement: a summary', *Inquiry*, 16(1): 95–100.

Neumayer E (2006) 'An empirical test of a neo-Malthusian theory of fertility change', *Population and Environment*, 27: 327–336.

Oreskes N and Conway EM (2014) *The collapse of western civilization: a view from the future*, Columbia University Press, New York.

Pantucci R and Ong K (2021) 'Persistence of right-wing extremism and terrorism in the west', *Counter Terrorist Trends and Analyses*, 13(1): 118–126.

Parr A (2018) 'Green scare', in Parr A (ed) *Birth of a new earth: the radical politics of environmentalism*, Columbia University Press, New York.

Pedroni L (2017) *Green vs. white: an examination of media portrayals of radical environmentalists and white supremacists* [master's thesis], San José State University, accessed 12 January 2023. https://doi.org/10.31979/etd.8m64-hgd4

Peet R, Robbins P and Watts M (2011) *Global political ecology*, Routledge, London.

Perry B and Scrivens R (2018) 'A climate for hate? An exploration of the right-wing extremist landscape in Canada', *Critical Criminology*, 27(1): 85–103.

Peucker M and Smith D (eds) (2019) *The far-right in contemporary Australia*, Springer, New York.

Polanyi K (1944) *The great transformation: the political and economic origins of our time*, Farrar & Rinehart, New York City.

Potter W (2011) *Green is the new red: an insider's account of a social movement under siege*, City Lights Books, San Francisco.

Raworth K (2017) *Doughnut economics: seven ways to think like a 21st-century economist*, Chelsea Green Publishing, Vermont.

Redclift M (2005) 'Sustainable development (1987–2005): an oxymoron comes of age', *Sustainable Development*, 13(4): 212–227.

Reynolds H (2006) *The other side of the frontier: Aboriginal resistance to the European invasion of Australia*, UNSW Press, Randwick.

Richards I (2023) 'Capturing the environment, security and development nexus: intergovernmental and NGO programming during the climate crisis', *Conflict, Security and Development*, https://doi.org/10.1080/14678802.2023.2211019.

Rose DB (2004) *Reports from a wild country: ethics for decolonisation*, UNSW Press, Randwick.

Ross AR (2017) *Against the fascist creep*, AK Press, Chico.

Ross AR and Bevensee E (2020) *Confronting the rise of eco-fascism means grappling with complex systems*, Centre for Analysis of the Radical Right (CARR), accessed 1 June 2022. https://www.radicalrightanalysis.com/2020/07/07/carr-research-insight-series-confronting -the-rise-of-eco-fascism-means-grappling-with-complex-systems/

Rueda D (2020) 'Neoecofascism: the example of the United States', *Journal for the Study of Radicalism*, 14(2): 95–125.

Ruddiman WF (2003) 'The anthropogenic greenhouse era began thousands of years ago', *Climatic Change*, 61(3): 261–293.

Sabin P (2013) *The bet: Paul Ehrlich, Julian Simon, and our gamble over Earth's future*, Yale University Press, New Haven.

Salter C (2011) 'Activism as terrorism: the green scare, radical environmentalism and governmentality', *Anarchist Developments in Cultural Studies*, 211–239.

Sassen S (2018) *The global city: strategic site, new frontier*, Springer Fachmedien Wiesbaden, New York.

Sasser J (2018) *On infertile ground: population control and women's rights in the era of climate change*, NYU Press, New York.

Schwartz P and Randall D (2003) *An abrupt climate change scenario and its implications for United States national security*, California Institute of Technology, accessed 19 March 2023. https://training.fema.gov/hiedu/docs/crr/catastrophe%20readiness%20and%20response %20-%20appendix%202%20-%20abrupt%20climate%20change.pdf

Shahbaz M, Bhattacharya M and Ahmed K (2017) 'CO2 emissions in Australia: economic and non-economic drivers in the long-run', *Applied Economics*, 49(13): 1273–1286.

Shiva V (2018) 'Earth democracy: sustainability, justice and peace', *Buffalo Environmental Law Journal*, 26(1): 1–15.

Slater L (2018) *Anxieties of belonging in settler colonialism: Australia, race and place*, Routledge, London.

Smith JK (2021) *The (re)emergence of eco-fascism: white-nationalism, sacrifice, and proto-fascism in the circulation of digital rhetoric in the ecological far-right* [PhD thesis], Baylor University, accessed 16 November 2022. https://baylor-ir.tdl.org/bitstream/ handle/2104/11491/SMITH-THESIS-2021.pdf

Spadaro PA (2020) 'Climate change, environmental terrorism, ecoterrorism and emerging threats', *Journal of Strategic Security*, 13(4): 58–80.

Staudenmaier P (2021) *Ecology contested: environmental politics between left and right*, New Compass Press, Porsgrunn.

Steffen W, Broadgate W, Deutsch L, Gaffney O and Ludwig C (2015) 'The trajectory of the Anthropocene: the great acceleration', *The Anthropocene Review*, 2(1): 81–98.

Steffen W, Persson Å, Deutsch L, Zalasiewicz J, Williams M, Richardson K, Crumley C, Crutzen P, Folke C, Gordon L and Molina M (2011) 'The Anthropocene: from global change to planetary stewardship', *Ambio*, 40: 739–761.

Swyngedouw E (2011) 'Depoliticized environments: the end of nature, climate change, and the post-political condition', *Royal Institute of Philosophy Supplements*, 69: 253–274.

Tayler L, Schulte C and Rall K (28 November 2019) 'Targeted: counterterrorism measures take aim at environmental activists', *Human Rights Watch*, accessed 1 June 2022. https://www.hrw.org/news/2019/11/28/targeted-counterterrorism-measures-take-aim -environmental-activists

Thomas C and Gosink E (2021) 'At the intersection of eco-crises, eco-anxiety, and political turbulence: a primer on twenty-first century ecofascism', *Perspectives on Global Development and Technology*, 20(1–2): 30–54.

van den Heuvel L (2020) *Who is (still) afraid of spectres haunting Europe? Comparing the concepts of 'Judeo-Bolshevism' and 'Cultural Marxism' in their respective notions of ecology* [master's thesis], Lunds Universitet, accessed 16 November 2022. https://lup .lub.lu.se/luur/download?func=downloadFile&recordOId=9029472&fileOId=9029473

Vanderheiden S (2008) 'Radical environmentalism in an age of antiterrorism', *Environmental Politics*, 17(2): 299–318.

Wainwright J and Mann G (2018) *Climate leviathan: a political theory of our planetary future*, Verso Books, Brooklyn.

Wallerstein I (1974) *The modern world-system I: capitalist agriculture and the origins of the European world-economy in the sixteenth century*, University of California Press, Berkeley.

Watson I (2014) *Aboriginal peoples, colonialism and international law: raw law*, Routledge, London.

Wright EO (2010) *Envisioning real utopias*, Verso Books, Brooklyn.

Zalasiewicz J, Williams M, Steffen W and Crutzen P (2010) 'The new world of the Anthropocene', *Environmental Science & Technology*, 44(7): 2228–2231.

Zerzan J (2002) *Running on emptiness: the pathology of civilization*, Feral House, Los Angeles.

Zimmerman ME (2004) 'Ecofascism: an enduring temptation', *Environmental Philosophy: from Animal Rights to Radical Ecology*, 4: 1–30.

2 A heated Australian landscape

Histories of environmental politics on the far right

Introduction

There has been a relative lack of organised, extra-institutional extreme-right political presence in Australia compared to other countries with European colonial histories (Moore 2005). This can be attributed in part to the country's geopolitical separation from continental Europe, the United Kingdom (UK) and the United States (US), but also to the ways in which mainstream political institutions in Australia have supported and readily absorbed white nationalist campaigns (Fleming and Mondon 2018). The mainstream connections to white nationalism in Australia reflect the country's colonial legacy, which includes the British genocide of Indigenous peoples who had been occupying the mainland for more than 65,000 years and extends to the current and ongoing criminalisation, dispossession and displacement of Indigenous peoples (Langton 2020; Rademaker 2022). Among other historical events and cases, the links to white nationalism also relate to the 1901 introduction of the Immigration Restriction Act, colloquially known – in combination with the 1901 Pacific Island Labourers Act – as the White Australia policy, which until its formal dissolution under Gough Whitlam's Australian Labor Party (ALP) government in 1973, prescribed almost exclusively European immigration to Australia (Piccini and Smith 2019).

Conspiratorial narratives of white demographic 'replacement' have persisted in dominant social and cultural institutions since colonisation in various ways (Hage 2012). These narratives reflect Australia's historical status as an isolated British outpost, with fears about threats posed by non-white populations first targeting Indigenous people and non-European migrants, then people from East Asia in particular, and more recently focusing on migrants from Muslim-majority countries (Cunneen et al. 1997; Elias et al. 2021). Conspiratorial expressions of demographic replacement in Australia have also connected with white genocide myths prevalent on the extreme right worldwide, ranging from the nineteenth-century antisemitic conspiracies of Judeo-Bolshevism (Hanebrink 2018), to current expressions of opposition to so-called Cultural Marxism (Busbridge et al. 2020) and the internationally prevalent Islamophobic register of French New Right thinker Renaud Camus's 'Great Replacement' theory (Obaidi et al. 2022).

DOI: 10.4324/9781003325437-2

Population replacement narratives have also played a prominent role in Australia's institutional–political history, as demonstrated by the rhetoric and actions of Pauline Hanson, the leader of the well-known far-right nativist One Nation Party (ONP). Hanson was elected to the Australian Parliament in 1996 on a Liberal Party ballot paper and again as an ONP Senator in 2018. In her first speech to the Australian Parliament, she declared: 'I believe we are in danger of being swamped by Asians' (Australian Human Rights Commission 2021). Her 1997 biography, *Pauline Hanson: the Truth*, elaborated claims of a plot developed by 'the internationalist elite of The New World Order', aimed at destroying Anglo-Saxon Australia through a process of 'immigrationism, multiculturalism, Asianisation, and Aboriginalism' (Hanson and Merritt 1997; Jones 2021). The book also mentioned 'the Aboriginal Question', signposting internationally resonant antisemitism tailored for a domestic audience, in explicit terms recalling the German Nazi regime and neo-Nazi propagandising about 'the Jewish Question', the answer to which was the Holocaust (Gordon 1984).[1]

Since the ONP's initial rise, far-right elements of nativist, antimigrant xenophobia have been echoed in other political projects, including billionaire mining tycoon Clive Palmer's United Australia Party (UAP). In the 2019 and 2022 Australian federal elections, Palmer and the UAP ran million-dollar advertising campaigns incorporating reactionary, Trumpian rhetoric based on conspiratorial messaging about supposed 'communist' threats to Australian conservative values, free enterprise and economic liberalism (Rimmer 2022). Extreme-right politicians such as Fraser Anning have also had a short foothold in the Australian political arena. After gaining seats running under the ONP banner, Anning served as an independent once in power. He then formed the Fraser Anning Conservative National Party and in his first Parliamentary speech in 2018 called for a 'final solution' to what he termed the Australian 'immigration problem' (SBS News 2018; WhiteRoseSociety 2019).[2]

Overt white nationalism has not only been endorsed in dominant institutions and at the highest levels in Australia. The 'mainstreaming' of white supremacist politics has also occurred through mechanisms of infiltration within Australia's two large traditional conservative parties, with state and federal National and Liberal parties having faced allegations of being targeted for infiltration by organised far-right and neo-Nazi groups (Seccombe 2018; Keane 2022). The connections between the political mainstream and the more extreme white supremacist fringe are, however, often more covert, discreet or ideologically based, as various chapters in this book seek to demonstrate. Rejecting atomised accounts of Australia's far-right political history, the book's investigation emphasises the porous ideological and material connections of influence evident in the use of environmental politics by disparate far- and extreme-right actors, both within and beyond political institutions.

The forthcoming account seeks to excavate the historical role of environmental politics for far-right actors in Australia, paying special attention to the use of environmental issues that also have contemporary resonance. This discussion illuminates patterns of ideological and propagandised responses to the Australian landscape, climate and natural environment that can be designated as patterns of quasi-fascist, nativist and other right-wing ideologies. Importantly, the investigation does not provide extensive analytical descriptors of the actors or individuals in question and is often partial in its account of their violent and supremacist politics. It also does not seek to identify spurious ideological coherence in what are often syncretic, selective and partial uses of environmental politics, deployed for the purposes of political persuasion (see Hughes et al. 2022). For clarity, the analysis speaks to the book's wider research agenda of identifying affiliations with respect to the environment among a spectrum of right-wing actors, aiming to do so without conflating disparate far-right elements.

In this chapter, four main topics are discussed. First, the intersection between Australia's genocidal colonial history and Australian myth-making is examined, particularly in the still-resonant origin narratives of a masculinist and extractivist Australian national identity based on differentiation from the British Empire, and the spiritual taming of the natural, local environment. Building upon this analysis, the subsequent section explores the enforcement of white nationalism in the early twentieth century across the mainstream political spectrum, including in early labourist–republican support for the exclusive rights of white workers and imperialist industrialists' rejection of egalitarianism by labelling socialist and labour movements as 'communist'. This also includes the emergence of fascist organisations such as the New Guard that opposed the redistributive economic policies of New South Wales (NSW) Labor Premier Jack Lang. This section highlights the white nationalist legacy evident in the establishment of Australian labour rights and the union movement, as well as the foundational implementation of the White Australia policy.

In the following section of this chapter, we discuss the environmentalist issues that motivated extreme right-wing groups and publications in Australia during the interwar period and in the opening stages of WWII. Drawing on David Bird's 2014 text *Nazi Dreamtime*, this section investigates how literary and cultural movements on the far and extreme right, which were sympathetic to the German Nazi regime at the time, included Odinist nativist writers and poets, as well as AFM supporters of nativist 'blood and soil' ideologies. The next section of this chapter examines the ruralist aspects of the ALR, which was the most well-resourced and longest-standing extreme-right organisation in the country (National Inquiry into Racist Violence 1991). The analysis investigates the organisation's involvement in the early post-war organic farming movement in Australia and its inability to present a radical alternative to conservative administrations' electoral victories in Australia (Moore 2005; Fleming and Mondon 2018). Brief attention is also paid to the influence of populationist thinking in Western environmentalism, and the discursive associations between environmental sustainability and anti-immigration sentiments in radical nationalist groups, such as the minor political party during the 1990s, Australians Against Further Immigration, and Jim Saleam's National Front formed in 1982. Finally, the discussion expands on the analysis of far-right actors' left-wing infiltration and substate orientation, focusing on expressions of bioregionalism, Third Positionism[3] and veteran military histories that were variously important to the territorial agendas of the ANM and NA groups from the 1980s to the 2000s.

Colonial genocide and origin narratives

Contemporary engagement with white nationalist environmental politics in Australia cannot be fully understood without considering the country's settler-colonial history, which was characterised by the systematic genocide of Aboriginal and Torres Strait Islander people. When Britain established the first penal colony in Australia at Botany Bay in Sydney in 1788, there were estimated to be over 750,000 Indigenous people in Australia. As members of the world's oldest continuous culture, they had been living in Australia for at least 65,000 years (Kiernan 2008; Langton 2020). Only approximately 60,000 Indigenous people remained by 1920, after enduring illness, disease and more than 400 systematic frontier massacres during colonisation (Attwood 2020).[4] In the period between 1910 and 1970, under assimilationist policies, an estimated 100,000 Indigenous children were also forcibly removed from their families and placed in adoption, foster care or children's homes, as part of the 'Stolen Generation' (Australian Human Rights Commission 1997; Funston and Herring 2016, 52). In the contemporary Australian context, ongoing state-based oppression resulted in more than 470 Indigenous people dying in custody by the 30th anniversary of the 1991 Royal Commission into Aboriginal Deaths in Custody (Human Rights Watch 2021).

Knowledge of the treatment of Indigenous people by Australian European settlers was effectively erased from many official accounts of the country's political history, including routinely in school curricula and in the media until recent decades (Moses 2004; Griffiths 2018). This historical amnesia has enabled the reimagining of the Australian landscape as *terra nullius*, an empty land available for settlement, and the justification of environmental policies that privilege settler interests over Indigenous ones (Lindqvist 2007; Hage 2017). In addition to the detrimental effects of colonisation on Indigenous people and their communities, another significant material impact of Australian settlement has been the suppression of knowledge about Indigenous people's relationships with the land, or 'Country', further contributing to the cultural and personal destruction caused by colonisation (Miller et al. 2010).

The lack of honest appraisal of Australian colonisation and its impacts has resulted in a lack of recognition of Indigenous peoples' spiritual and cultural practices based on land and their ecologically sustainable land management techniques, used for tens of thousands of years (McNamara 2014; Rickard 2017). These techniques represent an important point of juxtaposition against the unsustainable agrarian and industrial practices that underpinned early European settler-colonial empires, and which have now evolved into hyper-exploitation in energy extraction, animal agriculture, deforestation and the mining of essential minerals – all of which are primary contributors to global heating (Rose and Australian Heritage Commission 1996; Griffiths 2016; Sparrow 2022a). While these comparative methods and systems are not the main subject of investigation in this book, understanding their difference is crucial to understanding the related political and ecological impact of settler-colonialism in Australia.[5]

In understanding the enduring influence of settler-colonial mentalities, it is important to acknowledge how Australia's history as a British colony and penal colony for England had a profound impact on the settlers' relationship with both the British Empire and the local environment. This initial founding narrative contributed to the development of a unique Australian identity in the late nineteenth and early twentieth centuries, largely in representation of the challenges faced by working-class British and Irish migrant settlers (Elder 2020). This identity was characterised by a rejection of British colonialism and a desire for independence, centred around ideals such as self-reliance, independence and egalitarianism (Ward 1958; 1978), which were exclusively afforded to able-bodied heteronormative white men (Waling 2019).

Two key dimensions of a unique Australian national identity were the 'bushman' and the 'larrikin', which developed in opposition to British colonialism and in response to the trials of surviving in the rugged and unforgiving Australian landscape (Tranter and Donoghue 2007). While bushmen saw themselves as deeply connected to the land and valued physical labour, the larrikin rejected their British colonial masters – including those in the British command at Gallipoli and in the Eureka Stockade, and the Anglophile squatters in the 1890s – as well as women, who were seen as representatives of imperial fathers (Murrie 1998). Both identities celebrated symbolic self-reliance and independence, as well as physicality over intellectual pursuits (Ward 1958; Curthoys 1999), although different expressions of Australian national identities have either rejected individuality or promoted a 'rugged individualism' (Murrie 1998). The collective archetype of the Australian (white) man also rejected the materialism and consumerism of a modern industrial society, emphasising a simple way of life and a close connection to the natural environment. Interconnected mythic identities also emerged in opposition to the urban, industrialised societies of Europe and America, and each of these was a product of and a rebuttal to the British Empire.

Although the myths of national identity described above continue to influence the 'fair go' mentality of contemporary Australian political actors and inform who is included and excluded from the privileges associated with these identities, conflicting colonialist imagery and concepts are also routinely reinforced in mainstream political spaces alongside these founding myths (Bell 2014; Elder 2020). Colonialist themes and figures that are integrated into Australian political discourse express a reactionary form of national pride celebrating the conquest of the natural environment by European settlers and the subjugation of the Indigenous peoples who inhabited that environment (Tranter and Donoghue 2007). A primary example of mainstream Australian political institutions celebrating colonialism is the annual official 'Australia Day' ceremonies, parades and fireworks displays on 26 January, the date on which the British First Fleet arrived in Sydney in 1788. For many Indigenous people and their allies, the date represents the beginning of dispossession, violence and oppression; thus, its celebration is read as effectively dismissing or trivialising the ongoing harmful impacts of colonialism, and the day is often referred to as 'Invasion Day' or 'Survival Day' (Pearson and O'Neill 2009; Lipscombe et al. 2020).

Another significant element of the founding ideology for colonial Australia are romanticised identities rooted in the logics of capitalist resource exploitation, which at once reflect the dispossession of Indigenous peoples and inform prevalent current social attitudes towards environmental politics (Dalley 2022). Ideologically they relate to early colonisers' ability to exert control and ownership of territories, where resource extractivism was driven by capitalist accumulation incentives and the desire to derive profits from the land (Grégoire and Hatcher 2022). The early colonisers viewed the Australian landscape, and sometimes territories overseas, as a resource to be exploited for their own economic gain, rather than as a shared resource – or coexistent entity within the biosphere – requiring stewardship and care (Bedford et al. 2020; Walters 2023). This legacy of dispossession and exploitation has in the present day contributed to the development of a culture of 'extractivism', such that the relatively unfettered exploitation of natural resources has been constructed as a natural and inevitable part of the Australian economy (Harvey 1996; Hickel et al. 2022).

It is also important to note a related myth in the Australian context that portrays early colonisers as heroic figures who brought civilisation and progress to the 'wilderness', suggesting that they tamed the land and made it productive through their hard work and ingenuity (Harvey 2005). The concept of 'wilderness' as a pristine and untouched landscape is a romanticised falsehood that obscures the complex relations between humans and the natural world (Cronon 1996). Wilderness myths have roots in early European colonialism and were used to justify the subjugation and displacement of Indigenous peoples in the conquest of Australia, Canada, New Zealand, North America and elsewhere (see Rose 2011; Sparrow 2022b). In the contemporary context, these myths also reflect the racist dimensions of the mid-twentieth-century development of Western environmental thought and have been expressed to justify support for eliminationist population reduction policies (see Chapter 1, p. 4).

Eureka and White Australia

Unlike in some other national contexts, the varied narratives that emphasise 'white' conceptions of Australian nationhood have been characterised by competing colonial, federalist and republican elements since before Federation (Macintyre 2004).[6] While Aboriginal and Torres Strait Islander people were, almost immediately after colonial settlement, put into indentured labour (Curthoys and Moore 1995), from the 1850s, workers from diverse cultural and ethnic backgrounds were also routinely imported to work in agricultural and industrial labour, often in unfree and slave-like conditions (Atkinson 1991; Jayasuriya 2012). This exploitative labour system was closely tied to the enforcement white nationalist policies adopted across the political spectrum in the service of discrete economic objectives.

While the early organised labour movement in Australia advocated for better working conditions and full employment for British and Northern European workers, it also opposed the importation of non-white workers (Piccini and Smith 2019, 78). Fears of an influx of non-white migrants and the potential threat to the rights of white workers preceded the enactment of the Immigration Restriction Act in 1901, which together with the 1901 Pacific Island Labourers Act, which required the deportation of most Pacific Islanders beginning in 1906, established the colonial federation of states under the banner of 'White Australia' (Metcalfe 2013). Although employers had previously preferred cheap 'coloured labour', they also later supported the implementation of the White Australia policy to 'eliminate the reality of coloured entrepreneurial competition' (Burgmann 1985, 99).

The dismantling of the White Australia policy began in the 1940s, driven by the need for labour in Australia's rapidly industrialising economy. The Chifley Labor government (1945–1949) responded to this demand by allowing limited migration of non-Europeans, and this process was further expanded under the Menzies administration (1949–1966) from the 1950s (Smith 2020). The White Australia policy underwent a number of changes post-WWII, from 'assimilationism' (1947–1964), and 'integration' (1964–1972), to 'egalitarian multiculturalism' (1972–1975) (Jayasuriya 2012, 6). The policy though remained formally in place until its dissolution by the ALP Whitlam government in 1973 (Piccini and Smith 2019). Other 'multicultural' policies and social changes, though, such as the advent of 'liberal multiculturalism' (1975–1983) and 'managerial multiculturalism' (1983–2007), were variously characterised by a rise of divisive identity politics, conservative attacks on immigration, and an emphasis on 'productive diversity', insofar as multiculturalism centred around the dominant culture's 'universal' values (Jayasuriya 2012, 6–7).

It is also important to note that the legacy of the White Australia policy has persisted in various ways, even after its official abolition. Australia's immigration policies have remained heavily restrictive, particularly towards migrants from outside the Global North. The immigration system continues to privilege certain groups over others, with a strong emphasis on 'skilled' migrants who might benefit the Australian national economy (Ong 2006). This focus on economic contribution has led to an increasingly rigid system that often overlooks the human rights and well-being of migrants (Silverstein 2023). Moreover, Australia's policies towards irregular migrants, including refugees and asylum seekers, have been recognised internationally as harsh and in some cases 'torturous', with mandatory detention and offshore processing being key components of the country's approach (Laney et al. 2016; Hartley and Fleay 2017; see Chapter 6, p. 208).

In the present research, the racist context of White Australia is widely acknowledged, but the nativistic origins of republican Australianity, which developed partly through labourist agitation following the Gold Rush period in the mid-nineteenth century, are still often overlooked (McFadden 2023). This is particularly important for the subject of this book's investigation, since the legacy of early white workers' labour movements is currently reflected in the historicity of some contemporary extreme-right movements (see Chapter 3, p. 85).

During the Gold Rush, approximately 17,000 Chinese miners arrived in the Australian goldfields, and riots targeting them occurred across Victoria and NSW (Hu 2022). Australian and international 'diggers' rioted in response to the exorbitant licensing fees for the right to mine and own the gold they extracted, police brutality in collecting those fees, and a lack of representation of miners on the Legislative Council, which governed early mining activity (Connolly 1978). The racist aspect of several riots was rooted in the belief held by white miners that Chinese miners were unfairly taking their livelihoods. This sentiment was exacerbated by media and political speeches that portrayed Chinese migration, in particular, as a threat to white Australia (Markus 1985).

After several months of racist violence in Bendigo in 1854, one of the most famous events on the goldfields was the December 1854 Ballarat 'Eureka Rebellion', which culminated in miners swearing allegiance to the Eureka flag and protesting against the discriminatory practices of the colonial authorities, protesting through an armed rebellion and building a stockade to defend their protest and 'diggings' (Macintyre and Clark 2013). While the Eureka Rebellion did not have significant Chinese participation, and violence on that occasion was mostly directed towards the colonial authorities, a subsequent riot at Lambing Flat in NSW from 1860 to 1861 variously saw Chinese miners brutalised, their pigtails cut off, tools destroyed, and their camps, furniture and clothing set on fire, all under the banner of the Eureka flag (Kwok 2022).

Owing to its history, the Eureka flag (also known as the Southern Cross flag) has had contested meaning in Australia, being adapted as a symbol of working-class solidarity, workers' rights and democratic ideals into the political messaging of various labour groups, such as the ALP and many trade unions, including the Construction, Forestry, Maritime, Mining and Energy Union and the Australian Workers' Union (Sunter 2000; Clark 2018), as well as in a more limited fashion in the political messaging of some far-left groups who embrace its message of self-organisation, such as the Melbourne Anarchist Communist Group.[7] It has also, on the other hand, for some come to symbolise the exclusive rights of white workers in Australia to benefit from their labour (that is, 'white nationalism') and serves to communicate this message in contemporary far-right propaganda (MacDonald 2010; see Chapter 3, p. 89). The flag's importance in distinguishing different right-wing white supremacist tendencies rests largely on its identification with Australian nationalism as opposed to British colonialism (Mitropoulos 2018).

The Eureka flag's association with contemporary expressions of white supremacy was exemplified by its prominent display during the Cronulla 'riots' on 11 December 2005, when 5,000 people gathered near a Sydney beach to stage violent protests. At the event, white protestors carried placards with slogans like 'ethnic cleansing' and 'take Australia back', targeting people of Middle Eastern appearance and resulting in physical assaults (Johns 2008; Richards 2019). Illustrating its popularity among the extreme right, the Eureka flag was used as a symbol for the contemporary Australia First Party, founded by former ALP politician Graeme Campbell and currently led by Jim Saleam. Saleam was also a founder of the earlier neo-Nazi organisation, National Action in 1982, and a former member of the National Socialist Party of Australia (founded 1964) (Richards and Jones 2023; see also Chapter 2, p. 49).

The development of militant, anti-imperial and racially exclusive colonial claims to property rights and entitlement over the Australian landscape had a profound impact on the emergence of trade unionism and the formation of the ALP (Burgmann 1984). In a vivid demonstration of this, James Sinclair Taylor (JT) McGowen, the first Labor Premier of NSW (1910–1913), famously declared 'While Great Britain is behind us, and while her naval power is supreme, Australia will be what Australians want it – white, pure and industrially good' (Nairn 1969, 10). The Australian Workers' Union (AWU) played a particularly significant role in promoting racial exclusivity in Australian labour policies, supporting the White Australia policy and the implementation of its second phase in 1911 by Prime Minister Andrew Fisher. Although this example is just one among many others (Hunt 1978; Markey 1978; Crosby 2005), the second phase introduced an amendment granting 'absolute preference' to unionists while at the same time also supporting the dismissal of non-white workers (Martínez 1999). The exclusion of 'coloured aliens' from AWU membership did not extend to Indigenous people, who, in any case, were not granted the same rights as white workers (Martínez 1999). Before 1910, there was still significant employment of non-white labour in northern Australia, though this became controversial. At Port Darwin in 1910, then, the Department of Aboriginal Affairs supported Indigenous people working on the wharf being paid only two shillings per week with food and clothes, while white workers were demanding wages of two shillings per hour (Martínez 1999).

Although labour policies exclusively protecting white workers in Australia began to be dismantled in the early twentieth century, several unions continued advocating for strict immigration controls (Jupp 2002). These calls were often based on racist beliefs that increased migration to Australia would lead to higher unemployment rates and unfavourable bargaining conditions for white workers, resulting in wage cuts. However, this discriminatory attitude was not consistent among all members throughout the historical evolution of the labour movement, and other workers' movements did not uphold such views. The Industrial Workers of the World (IWW), a radical union, significantly impacted the Australian labour movement while advocating for unity among all workers, regardless of their background, viewing migrants as allies in the fight for better working conditions (Burgmann 1984, 42). The Communist Party of Australia (CPA) was also a vocal opponent of the White Australia policy during the post-WWII era, guided by their internationalist Marxist analysis of imperialism and the belief that working-class solidarity should transcend national and racial boundaries (Piccini and Smith 2019).[8] Student activist groups of the Australian 'New Left' in the 1960s emerging from within the Australian labour movement and affiliated with the ALP also eventually drove its abolition (Piccini and Smith 2019).

While white nationalist politics were significant in the formation of the early labour union movement in Australia, it is also important to recognise how other forms of white supremacism constituted a significant part of the far- and extreme-right social reaction against organised workers, which had been particularly accelerating through the early twentieth century (Cathcart 1988). In the lead up to WWII, reactionary groups in Australia promoted a social backlash against union organising, opposing the egalitarian values of a progressive left, as well as 'communism' and international socialism (Burgmann 2016). In Australia, as internationally, and through history, these tendencies were motivated by the long-held desires of powerful figures with commercial and industrial interests to maintain their wealth and power, which they saw as threatened by unions and other workers' collectives (Fleming et al. 2004; Zinn 2015).

One prominent industrialist on the far right was Colonel Eric Campbell, who led the New Guard, a fascist paramilitary group with more than 50,000 members (Smith et al. 2023a, 3) that emerged in 1931 to oppose the NSW Premier Jack Lang, a popular politician revered by his supporters as 'greater than Lenin' (Sparrow 2015). The violent core of the New Guard, the Fascist Legion, comprised militias targeting unionists and seeking to physically attack official figures in the labour movement, while members believed that they were saving Australia from a 'communist takeover' (Sparrow 2015). Another leader of the New Guard, Captain Francis De Groot, stated in reflecting upon the period, 'we had no intention of handing over Australia to the tender mercies of the rubbishy kind of people who aspired to rule us' (Moore 2005, 7). Members of the New Guard also expressed support for Nazi Germany, and these events were occurring at the same time as the Nazi Party's banning of trade unions, and its first deployment of paramilitaries to disrupt workers' meetings and intimidate union officials (Adamson 1980). Campbell, in his own words, reflected on the period as follows:

> It was the likelihood of industrial unrest that gave me concern. A series of strikes inspired by the Communists and encouraged by the Trades Hall as a 'softening up' for Socialism could have paralysed the city in a matter of hours, and for want of food and water the whole population could well have been at the mercy of a determined minority ... The State government had been elected only a matter of months before with a large majority. The plank of socialization of production, distribution, and exchange had been a basic one in Labor's platform for as long as I can remember.
>
> (1965, 60, 84)

Despite the New Guard's most infamous act being Captain De Groot's disruption of the opening ceremony of the Sydney Harbour Bridge to humiliate Premier Lang on 19 March 1932 (Moore 2005, 12), it is important to acknowledge that the political ideology it espoused reflected broader white supremacist, conservative and imperialist tendencies that were more openly expressed before WWII. The National Civic Council (NCC), which evolved in 1957 from the Catholic Social Studies Movement founded in the early 1940s by BA 'Bob' Santamaria, also vehemently opposed not only the CPA but also trade unions and the labour movement (Considine 1985). While the NCC's opposition to the labour movement was rooted in its belief that unions were inherently communist and aimed to subvert the 'Australian way of life', the organisation promoted Catholic social teaching, which prioritised individual responsibility and property rights over collective action and workers' rights (see e.g. Henderson 2015). Civil associations and publications led by influential businessmen also supported anti-union and anti-labour agitation. For example, media mogul Keith Murdoch, father of News Corp Chief Executive Officer (CEO),

Rupert Murdoch, used his own media empire to propagandise against left-wing politics and labourism. Although Keith Murdoch supported British imperialism and opposed Nazi Germany, his views were notorious for shifting over time, also having controversially expressed support for the Italian fascist leader, Benito Mussolini (McKnight and McChesney 2013).

Blood and soil, Odinism and 'Aryan Indigeneity'

Australia's colonial legacy of exploiting local lands and territories for profit, often at the expense of Indigenous peoples, continues to this day through large-scale resource projects that result in displacement and dispossession. However, some far- and extreme-right actors in the lead up to WWII were less concerned with sustaining patterns of domination and extraction, instead aiming to establish a radical nativist movement based on the ideals of 'race and place' (Stephensen 1935; Slater 2018). As WWII approached, new figures, publications and organisations emerged, creating connections between far-right and extreme-right ideology in Australia and environmental politics. An emerging radical nativist cultural and literary movement was influenced by quasi-*völkisch* sentiments of 'blood and soil' imported from Nazi Germany during the interwar period (Moore 1989). David Bird's 2014 book, *Nazi Dreamtime*, offers insights into this era. The brief summary provided below, drawing upon that text, explores the actions and influence of figures in Australia who were attracted to Nazism before and after Hitler's rise to power in 1933.

Influential political figures in Australia expressing admiration for national socialist politics leading up to WWII included writers, politicians and poets who developed connections with Nazi officials and embraced national socialist political ideals (Munro 1984). Through literary magazines, conferences and news media constituting a form of *Kulturkampf*, referring to 'cultural struggle' – or a Gramscian form of metapolitics (Teitelbaum 2019) – they promoted the development of a unique Australian form of white nativist identity predicated on associations between culture, race, land and territory (Bird 2014). One way in which these nationalist actors sought to popularise their ideologies was by gradually incorporating poets, artists and prose writers such as Xavier Herbert, Miles Franklin and Eleanor Dark into the forerunners to the AFM project led by Percy Reginald (PR) 'Inky' Stephensen (Macklin 2020). Stephensen, a writer, activist and influential publisher of the 1920s and 1930s, was sometimes colloquially referred to as the 'Bunyip critic'. After he established the AFM in 1941, some such as Herbert and Franklin had already left the movement. Other early contributors to AFM though included Adela Pankhurst, the youngest daughter of the prominent British suffragette Emmeline Pankhurst, Melbourne activist Leslie Cahill and barrister and author Alexander Rud Mills (Bird 2014, 317). Early developments within the organisation were also influenced by the 'Australia First' election campaign of businessman William John Miles in 1943. Miles himself published the Nietzschean, proto-Nazi newsletter the *Publicist* from 1936 to 1942, after which time Stephensen took over the publication (Bird 2014, 323).

The AFM was a fascist organisation that sought to promote a white, Anglo-Saxon identity and exclude non-white and non-Anglo-Saxon peoples from Australian society. With several of its members arrested or interned during WWII, the organisation was characterised by its admiration for Adolf Hitler and its advocacy of totalitarianism and racial purity (Munro 1984). The AFM held antisemitic, anti-imperial and anti-socialist views and self-described as in favour of 'national socialism' (Macklin 2020). Importantly, the *Publicist* served as a key mouthpiece for the AFM and existed as part of a larger ideological framework of ultraconservative and far-right outlets in Australia during that time, including the *Catholic Weekly*, *The Advocate*, Jack Lindsay and John Kirtley's *Fanfrolico Press* and the *National Socialist* (Bird 2014).

In his writings, Stephensen drew upon and incorporated ideological tropes that were already present in major literary works of his time, including DH Lawrence's focus on vitality and antimodernity and Miles Franklin's depictions of the Australian outback. In 1932, Stephensen had established Endeavour Press in Sydney with Norman Lindsay, which produced over a dozen titles by writers such as Banjo Patterson and Miles Franklin. Historically, Stephensen also had claimed communist ideological influence but later declared himself an 'Australian Nationalist Socialist', drawing inspiration for his 1936 text, *The Foundations of Culture in Australia*, from the Nietzschean literary tendencies of Jack Lindsay and William Baylebridge, the latter of whom also authored an unpublished poem in 1913 titled 'Palingenesis' (Bird 2014, 147). In *Foundations* Stephensen described:

> What then of culture in Australia? Here is not a mere vicinity, but a whole continent, unique in its natural features, and unique in the fact of its continental homogeneity of race and language. Australia is the only continent on the earth inhabited by one race, under one government, speaking one language. The population at present is not much greater than was that of Britain in Shakespeare's time, but by the end of the twentieth century we may expect that the population will expand to at least twenty millions [*sic*], remaining of European parent-stock, but with locally-developed characteristics, and with a locally-created culture. Australia will then become indubitably recognised as a nation, and will lose all trace of colonial status.
>
> (Stephensen 1935)

Through *Foundations*, the *Australia First Manifesto* and other texts, Stephensen sought to establish a mythical eugenicist basis for 'blood and soil' in the Australian native context (Encel 1989). He also did this in part through a spurious thesis asserting a connection between the genetic bloodlines of Indigenous people in Australia and his own 'Aryan' race (Munro 1984, 208). Bird cites the probable origins of this theory as the Black Caucasian thesis[9] expounded in 1925 by Adelaide-based anthropologist Herbert Basedow (Bird 2014, 71). To these ends, the editors of the *Publicist* and founders of the AFM courted members of the Jindyworobak Club, a nationalist literary movement founded in 1937 at the University of Adelaide. The club was founded by Rex Ingamells, an Australian poet, novelist and teacher, in

1937. Stephensen claimed to Ingamells that Indigenous Australians were '"the oldest Aryans on earth" ... the "Ancient Aryans" of the aboriginal race' (Bird 2014, 72).

The oppression of Indigenous peoples by the nativist extreme right was compounded by the fact that several prominent nationalist literary actors in the 1920s and 1930s in Australia expressed superficial nominal support for the plight of Indigenous peoples while exploitatively drawing aesthetic inspiration from both the 'German dream' and the Aboriginal Dreamtime, or *Alcheringa*, as a collective model for an Australian nativist ideal (Bird 2014, xii). Members of the Jindyworobak group and AFM proponents appropriated Indigenous symbols and cultural heritage, blending their aesthetic aspects with quasi-Nietzschean expressions characteristic of *völkisch* thought. They often also emphasised Aryan or Nordic associations with the concept of being 'white' (Bird 2014).

The Jindyworobaks, or 'Jindys', published their manifesto *Conditional Culture* in 1938, borrowing the 'Jindyworobak' name from James Devaney's *The vanished tribes* where it had been appropriated from the Woiwurrung language, meaning 'to annex, to join', with Devaney himself also becoming a Jindy (Bird 2014, 165). While Jindy publications were scattered with Indigenous imagery, terminology and references to Australian fauna and flora, they assimilated Indigenous culture into a white nativist framework emphasising a 'pristine outlook on life', along with the importance of land and place connection (Bird 2014, 166). A 1940 *Corroboree* collection of poems by Australia-First writer and Jindy proponent Ian Mudie, featuring a dedication by Stephensen, displayed on its cover an Indigenous man dancing before a blazing sun. The collection was an agitation against technological modernity in a spirit Mudie described as 'men affirming manhood, women / affirming womanhood, nations / daring to affirm their nationhood', replacing a world in which men and women were 'economic puppets of the new calf' and 'automata of the bright machines' (Bird 2014, 282). Despite the Jindy club attracting members during the 1940s such as Jack Lindsay, a British-born writer and scholar who moved to Australia in the 1930s and wrote positively about Nazi Germany, and the ultranationalist Australian poets John Kirtley and Alexander Rud Mills who were both interned during the war, it continued to appeal to a wide audience, publishing an anthology with 160 contributors in 1948 (Bird 2014).[10]

The far-right aspects of the nationalist literary movement in Australia in the 1930s and early 1940s, were importantly characterised by an ideological emphasis on Evolian spiritual racism, or Odinist or Nordic religion, more than they were focused on either a workers' paradise or Indigenous folklore (Winter 2005). Among the several influential contributors to the nationalist literary movement was Melbourne solicitor Alexander Rud Mills, also known as 'Tasman Forth', who acted as a prominent Odinist mystic and core member of the AFM, and who had also met Hitler in Munich in 1933 (Henderson 2005; Macklin 2020). In his 1933 book *Fear*, Mills outlined how 'the new reality would emerge based on "instinct" and the service of those things which appealed to the *völkisch* mind – family, state, nation, and race' (Bird 2014, 121) – demonstrating a Nietzschean perspectivist influence that could perhaps be said to resonate with contemporary post-truth politics (see Beiner 2018). Mills also believed that 'Odinists, and the Odinists only, will preserve our race', and, as Bird describes, 'bind the 70 million white "British" of the homeland and the dominions together against "the Jew and his outlook"' (Bird 2014, 125). He promoted the *Protocols of the Elders of Zion* and held meetings at a site in the Dandenong ranges in Victoria, Australia, called the Brown House, which was imitative of the Nazi Palace of Obersalzberg in Munich (Bird 2014, 118).

Another leading figure of the AFM, William Hardy Wilson, shared the admiration for Japan and the Nazis with Stephensen. In his 1929 work, *The Dawn of a New Civilisation*, and a 1936 sequel called *Collapse of Civilisation*, Wilson expressed his own esoteric quasi-Eurasianist position, arguing that the cold climates of Europe would lead to a lack of creativity and ultimately result in civilisational decline (see e.g., Shekhovtsov 2017), believing that the warm temperatures in Australia would lead to an 'Asian renewal' to begin in Japan (Bird 2014, 132). Wilson also proposed plans to establish concentration camps in the Dandenongs to accommodate Jewish refugees arriving in Australia from Germany. He suggested creating a settlement called 'Israelia' in the hills near Wandin, a concept recognised at the time as racist due to its segregating and securitising aspects, with the plan including the establishment of watchtowers on Mount Dandenong and Mount Donna Buang (Bird 2014, 134).

Prior to WWII, the Australian far right's divergence in ideology between a focus on 'Aryanism' versus 'Nordicism' could in some ways be said to mirror a similar ideological division in the German Nazi regime. Himmler's mystical faction preferred Nordic religion, while General Erich Ludendorff was an Odinist, and Nazi theorist and ideologue Alfred Rosenberg was a follower of the Aryan Christ. In contrast, Hitler's adherents considered the Aryan race itself to be divine (Bird 2014, 114). The development of a race-based Australian nativist identity through articles published in far-right literary outlets during the 1930s and 1940s, also showcased an intermingling of different ideological perspectives. In Mills's *Publicist* articles, for example, Bird highlights a blend of paganism and Australian national socialism, eventually in post-1940 publications featuring the Indigenous veneer of Jindy literature (2014, 256).

Bird's (2014) analysis of artefacts from the period highlights the prevalence of pseudo-environmentalist ideologies in early WWII-era homegrown national socialism, with the idea of foreign 'invaders' taking over native lands occupied by both old and new 'Aryans' being a common theme. This sentiment extended beyond literary circles, to the media and academia. For example, Augustin Lodewyckx, an academic known for his support of Nazi and Fascist ideologies, engaged his students in 'quasi-Wondervögel outdoor activities' like hiking in the Australian mountains to instil nationalistic values (Bird 2014, 207). In an article he wrote for the *Argus* journal on Hitler's 46th birthday, Lodewyckx proposed using Nazi Germany and Fascist Italy as models for Australia's demographic reconfiguration, citing the falling birth rate of the white race as a concern (Bird 2014, 203). He also expressed agreement with Professor Alan Chisholm, another influential linguist of the University of Melbourne, and a Nazi sympathiser, who advocated the perspective that representative democracy's 'majority rule' amounted to 'mediocrity rule' (Bird 2014, 184 196).

Ruralism, the organic movement and the Australian League of Rights

During the Cold War era, there was a growing preoccupation with globally connected communist movements. This concern was particularly prevalent among the far and extreme right, especially in Anglophone countries with European colonial histories (Piccini et al. 2018; Smith 2020). The notion of 'White Australia' was viewed during this period as part of a global struggle against the perceived encroachment of liberalism in the post-war era, as well as against the adoption of multiculturalism or 'multiracial democracy' (Smith 2020, 231). On the far right, sentiments against immigration and populationism were also often rooted in opposition to the gradual relaxation of immigration controls occurring in the 1950s under the Menzies administration (Lowe 1999; Smith 2020), in the context of what is sometimes referred to as the 'post-war immigration boom' (Collins 2008).

Following WWII, the growth of cities due to industrialisation, along with the emergence of multicultural Australia, contributed to a renewed focus on anti-urban sentiments among the far right (Jayasuriya 2012; O'Carroll 2020). These attitudes often incorporated nostalgic myths of traditional and conservative rural life (Smith et al. 2023b) and blamed urban elites and migrants for perceived societal decline and environmental degradation (Wilson 1992; Forchtner and Kølvraa 2015). Fearmongering tactics were used to stoke resentment, including conspiracy theories about 'no-go zones' controlled by people from marginalised non-white groups (Rudner 2019). Although progressive tendencies certainly existed among regional populations, far-right politics in some rural communities developed from the socially conservative ideological foundations of the Country Party, which was formed in 1920, later becoming the National Country Party in 1975 and the National Party of Australia in 1982 (Graham 1966; Davison and Brodie 2005).

One group exemplifying these tendencies, though antisemitic and avowedly racist where the Country Party was not, was the ALR, the longest-standing extreme-right movement in Australia's history (Campbell 1978; Gaynor 2012). The ALR movement was largely concentrated within rural Australia, across the 'Darling Downs and Kingaroy in Queensland, Cobar and Bourke in New South Wales, the Eyre Peninsula in South Australia, and the electorates of Moore, Forrest and Canning in Western Australia' and 'in Colac, the Murray electorate, Shepparton, and parts of Melbourne including Frankston and Diamond Valley-Eltham' (Greason 1997, 193). Eric Butler, a WWII veteran, founded the organisation in South Australia, in 1946 (Moore 2005), in the same year publishing a condensed version of the *Protocols of the Elders of Zion*, titled *The International Jew* (Greason 1997; Mendes 2022). In political speeches during the 1960s and 1970s Butler praised the white minority regimes of Rhodesia and South Africa as necessary to maintain order and prevent the spread of communism in Africa (Leibler 1967; LeCras 2023). Despite its extreme right-wing ideology, the ALR had several connections to mainstream political institutions and ran concerted campaigns to enter politics.

The ALR's platform included extreme British Empire loyalism (or 'monarchism') and anti-communism, with members infiltrating the Australian Country and Liberal parties via Butler's affiliation with the UK League of Empire Loyalists (Smith 2020; LeCras 2023). The organisation also established front groups through which to recruit and propagandise, such as the Institute for Economic Democracy, the Australian Heritage Society, Ladies in Line Against Communism and the Union of Farmers, and it controlled or had significant influence over the conservative and nationalist presses, *Australian Conservative* and *New Times* (Gaynor 2012; Greason 1997, 196; National Inquiry into Racist Violence 1991, 201; e.g. Butler 1951, 1952). Prominent ALR organisers, including Arthur Chresby, served as a bridge between extreme-right activists and professional political institutions (Smith 2020). For example, the ALR enjoyed support from corrupt former Queensland Premier Sir Joh Bjelke-Petersen, expressed solidarity with Pauline Hanson after her 1996 parliamentary speech and had links to Graeme Campbell after his departure from the ALP (Greason 1997; Ashton and Cornwall 2006). The organisation also courted mainstream politicians, including former Liberal Party Senator (and Opposition Leader) Alexander Downer, who spoke at the annual state seminar of the ALR in South Australia in 1987 (Greason 1997, 197).

The 1991 National Inquiry into Racist Violence referred to the ALR as 'the best organised and most substantially financed, racist organisation in Australia', further proclaiming that 'it's resources, influence, stability and professionalism far exceed those of any other racist organisation in Australia, past or present' (National Inquiry into Racist Violence 1991, 200). Butler and the ALR did not have an interest in developing a nativist identity by superficially appropriating Indigenous imagery or ideas, as did the AFM. Instead, they advocated extreme British Empire loyalism, emphasising strong allegiance to the British monarch, as well as a belief in the importance of maintaining the British Empire and the Commonwealth (Henderson 2005). Like the AFM, though, the ALR was antisemitic, spreading Butler's conspiratorial view that a Jewish-controlled economic 'New International Order' was behind the Indigenous land rights movement, as well as the destruction of rural farms, and civic movements towards 'multi-racialism' and 'multiculturalism' (Greason 1997; Markus 2018).

The ALR was nationalised in 1960. In the 1970s, the group released pamphlets focusing on organicism, the merits of rural farming, and traditionalist familial life, such as 'Ecology and Us' (1971). David Greason, a former member of the ALR and author of the 1994 book *I Was a Teenage Fascist*, outlined the ALR's ruralism and radical anti-urbanism in a chapter titled 'Australia's racist far right', featured in Cunneen and their co-authors' edited book *Faces of Hate* (1997). In that chapter, Greason stated that the ALR believed the land rights movement was 'a communist plot to establish an independent, armed and black republic within Australia' (1997, 193), and elaborated that ALR's values of empire loyalism, social conservatism and Christianity were promoted largely through its monthly newsletter, *Intelligence Survey*, and through leafletting and bookshop distributions (Greason 1997). In a retrospective defence of the ALR, a 1985 text titled *The Truth about the Australian League of Rights* written by Butler himself referred to the importance of rural values, land-based spirituality and rejection of the features of the financial or economic system characterised by government intervention and control:

> When debt finance, crushing taxation, monetary inflation and centralised power destroyed the independence of the sturdy Roman peasantry, the backbone of the Roman Empire was broken. Stalin knew he could never impose complete totalitarianism until he destroyed the Russian farmers. The future of Australia requires an independent rural Australia. I have always been struck in considering the development of Christian influences on the English social structure, that England combined a tradition of liberty in close association with a deep passion for the land. It was the great lover of rural England, William Cobbett of 1st century [*sic*], a man who attacked the debt system, who drew attention to the philosophic cleavage which resulted in the surrender of traditional Christian standards to pure economics. The new philosophy divided God from nature and man from both.
>
> (1985, 109)

The ALR's ruralist attitudes extended to its attempted infiltration of the early organic movement, which was nominally progressive though influenced by various ideologies, including political developments in England. In 1946, for instance, after the ALR established its Victorian branch, Eric Butler and John Weller, an ALR member who became an editor of the *New Times* in 1946, sought to directly recruit from the local organic community. Butler also delivered a speech to the Victorian Compost Society about his organic farm in Panton Hill in 1951, and his talk was reproduced in *New Times* magazine and *Victorian Compose News* (Gaynor 2012, 262). To illuminate the wider context in which these events occurred, it is important to note that the roots of Australia's organic movement from this period lie in the UK Soil Association, which was co-founded by Viscount Lymington, who had links to Nazi Germany. The association also included Jorian Jenks of the British Union of Fascists as a member from its inception in 1946 until his death in 1963 (Coupland 2016; Gaynor 2012; Paull 2013). Additionally, the Australian *New Times*, a one-time progressive alternative news magazine, shared similarities with the *New English Weekly* newspaper, both promoting CH Douglas's social credit theory (discussed below) and featuring articles on topics such as rural affairs, health and farming, often within the framework of Christian social thought (Gaynor 2012).

The early organic movement in Australia was concerned with issues of soil erosion and dust storms that had plagued the country for decades (Jones 2008). These concerns were linked to a broader belief that the state of the soil was intimately tied to the fate of the nation, race or civilisation (Gaynor 2012). It was only in the late twentieth century that the Australian organic movement shifted its focus to opposing the genetic modification of food, promoting sustainable agricultural practices and encouraging healthy eating (Gaynor 2012). The early movement's politics and ideas were disseminated through gardening magazines, journals and nature writing, and it had ties to a range of political groups, from the progressivist left to the far right (Gaynor 2012; Head 2022). The importance of soil for both social and agricultural well-being was at the time a unifying theme for many adherents, with some on the far right attracted to the movement's related focus on 'natural law' and rejection of both industrialism and modernity (Jones and Gaynor 2019; see also Boukala and Tountasaki 2019). While the organic movement was also associated with recreational activities such as small-scale farming and bushwalking, protecting the health of soil was seen as a critical link to local communities and a defence against the 'degenerative' processes of industrialisation and urbanisation (Gaynor 2012; see also Moore and Roberts 2022).

Social credit theory, with which the organic movement was closely associated, posited that economic crises were caused by a disparity between wages and prices, and that during such times, purchasing power inevitably declines. In response, social credit theoretician CH Douglas proposed a solution in the form of debt-free cash payments by governments to workers or producers who sold their goods for less than market rates (Wentworth 1931; Pullen and Smith 1997). While this theory had supporters across the political spectrum, it also attracted criticism for its association with antisemitic conspiracy theories due to its characterisation of the international financial system. In Australia, for instance, the *New Times* promoted social credit monetary policy to an organic farming audience, but it also published *The Protocols of the Elders of Zion* in its inaugural year of publication in 1934 (Gaynor 2012). Social credit had been central to the ALR, with Butler contributing to the *New Times* publication in 1937. Butler's adherence to the concept of the natural 'law of return' (discussed below) also earned him appreciation from Henry Shoobridge, the founder of the Living Soil Association of Tasmania. John Weller, an ALR member, became the editor of *New Times* in 1947 (Gaynor 2012, 262).

During the late 1960s and 1970s, then, Western environmentalism emerged, with some elements promoting erroneous claims that overpopulation in the Global South was the root cause of environmental degradation (Moore 2016). While the natural law of return in agriculture refers to the process of nutrients and organic matter being recycled back into the soil through natural decay and decomposition, the far right in Australia incorporated this concept into their propaganda messaging and advocacy of anti-immigration policies to promote the genocidal principle of creating racially pure societies free from external influences (Merrill 1983; see e.g., the comments of Arthur Calwell, leader of the ALP from 1960 to 1967, Tavan 2012).

Calls for population reduction measures based on concerns about pollution, global warming and resource scarcity were exploited by the far right in Australia, as they were in other countries. The ALR, for instance, championed the idea of 'racial purity', claiming that non-white immigration would harm the Australian environment and undermine the 'Australian' way of life (Richmond 1977; Garton 2010). It is also important to note that the influence of Western environmentalism on the development of far-right ecological thought extends beyond Australia, as discussed in other chapters of this book (see e.g. Chapter 3, p. 72, Chapter 5 p. 187, Chapter 6, p. 210).

The emergence of organisations such as Australians Against Immigration, formed in 1988 by a respected middle-class physician, Dr Robert Spencer, and his wife Robyn Spencer – which then became Australians Against Further Immigration (AAFI) in 1989 – was also influential. With reference to the AAFI, Jim Saleam, at the time the former member of the late 1960s National Socialist Party of Australia and future co-founder of the AFM, described the importance of environmentalist ideologies to the far right in his PhD thesis as follows:

> The synthesis of an environmentalist argument against immigration with a populist-nationalism directed at overseas economic forces and internal traitors (to whom liberal intellectuals were soon added) was inherently radical.
>
> (1999, 312)

While the AAFI was joined by other conservative groups such as Australians for an Ecologically Sustainable Population and Writers for an Ecologically Sustainable Population, Saleam himself also made his own attempted incursions into the populationist, pseudo-environmentalist far-right debate (Saleam 1999, 311). He founded the National Alliance in Sydney with far-right intellectuals Frank Salter and Eddy Azzopardi in 1978, and this group went on a few years later to merge with the Progressive Conservative Party and the Immigration Control Association, forming the Progressive Nationalist Party (Smith, Persian, and Fox 2023a, 6). The Strasserite,[11] Third Positionist anti-capitalism of that group also underwent further ideological development in 1982 when Saleam and his partners formed National Action, a group that predominated on the Australian far-right scene in the 1980s and 1990s (Henderson 2005; Whitford 2011). NA was responsible for street attacks such as firebombings, racist 'white pride' marches and physical attacks, one of which resulted in Saleam's three-year incarceration for attacking the Sydney home of Australian National Congress representative Eddie Funde in 1989 (Whitford 2013; Smith et al. 2023a, 7).

Militarism and bioregionalism

During the 1980s and 1990s, another new form of pseudo-environmentalist, territorial politics emerged on the Australian far and extreme right. This ideology was influenced by more complex bioregionalist political philosophies, which were growing in prevalence among progressivist groups at the time, but which were often superficially expressed by the far right as support for racially homogeneous 'folk communities' (Sewell 1995; see e.g. ideas discussed in Jim Saleam's PhD thesis; Saleam 1999). Far-right actors and movements that adopted certain aspects of bioregionalist philosophy emphasised the importance of defining human communities based on the natural ecological boundaries and systems of a particular region. However, they usually engaged with this perspective only to the extent that it aligned with their principles of 'natural law' and natural hierarchy (Lubarda 2020). In this context, blood and soil ideologies translated into the bioregionalist idea that ecological factors such as climate, geography, soil and water create unique regions with their own distinct ecosystems, cultures and social systems (Sale 2000; Taylor 2019).

Another aspect of nationalist politics that gained traction, particularly from the 1980s, was the recruitment of radical right groups from veteran communities and the adoption of their ultranationalist valorisation of military service (Kulmar and Jensen 2021). Although not strictly environmentalist, this development in far-right politics in Australia often involved symbolic messaging about protecting the Australian 'homeland' through associations with the 'Anzac legend', diggers, bushman national identity and culture, and (sometimes reluctantly) loyalty to the British Empire (Peucker et al. 2021; Smith and Iner 2021).

Recognising the significant role that militaristic ideologies have played in the development of far-right and extreme-right social movements is crucial in this context. Although the majority of veterans and defence force personnel are not of the far right, military ideologies are amenable to actors who would exploit martial politics, authoritarian tendencies and patriotic values, not only in Australia but also worldwide (Smith et al. 2023b). Although the importance of militarism to the far right is often overlooked in analyses of these movements, military service, training and ideology have been vital to many Australian far-right figures, past and present (Nilan 2019; Gillespie 2020), with the leaders of most Australian twentieth-century white supremacist organisations mentioned in this chapter having histories of military service.

White supremacist recruitment from the Returned and Services League of Australia (RSL) – a coalition of Australian military veterans – also became a significant issue beginning in the 1980s, with veterans who struggled with the transition to civilian life finding a sense of belonging and purpose in far-right groups that promoted a militant form of nationalism (Kulmar and Jensen 2021). In this context, the term 'digger' carries a secondary meaning beyond its original gold-mining connotations. It refers to the heroic status attributed to Australian and New Zealand soldiers who, while sometimes associated with miners in the gold fields, also dug trenches on the Western Front during WWI, symbolising the 'Anzac legend' (Peucker et al. 2021). Deeply ingrained in Australian culture through popular songs, films and literature, the Anzac legend was partly constructed through political and literary discourse, as well as by figures such as historian Manning Clark, to shift the emphasis from glorifying war and focusing more on the human aspects of the soldiers' experiences (McKenna and Ward 2007). It has also though been argued to promote a narrow version of national identity and a conservative political and cultural agenda, prioritising war over peace and reinforcing conservative social and political values (Lake and Reynolds 2010).

As discussed earlier, Jack van Tongeren, a Vietnam veteran known for his vehement anti-communist stance, led the ANM, a neo-Nazi group that embraced sub-state ethno-communities, emphasised militarism and had around 100 members in the mid-1980s. The ANM drew support from a limited but notable contingent of RSLs. In some ways, the ANM's anti-industrial and anti-technological stance echoed the endorsement of the 'laws of nature' by the ALR and the early Nietzschean emphasis of the AFM on the natural development of 'instincts' (Bird 2014, 121). Van Tongeren also evoked superficial bioregionalist-inspired ideas of localisation, describing the organisation in *The ANM Story* as advocating for 'racially pure folk communities, bound together by loyalty to the Blood and Soil and living culturally fulfilled lives in harmony with the rest of nature' (cited in Campion 2021, 13).

The ANM drew inspiration from various historical figures, such as Ned Kelly, Henry Lawson, and a group of poets, miners and bushmen antifascists have described as a 'proto-fascist historical cadre' opposed to the British colonial establishment and its attempt to 'Asianise' Australia through the importation of workers (Fox 2022). However, the reality of the ANM's activities was even more disturbing. From 1984 to 1989, the ANM launched a half-million-poster campaign in Perth, targeting Asians, people of colour and Jewish communities, and this escalated to localised incidents of extreme racist violence (Greason 1997; James 2005). ANM members firebombed Chinese restaurants and committed armed robberies to finance their attacks. Jack van Tongeren was fined AUD1,700 for possessing explosives and unlicensed ammunition, and he eventually served 13 years in prison for arson (Fox 2022). After his release in 2002, Van Tongeren led the ANM's successor organisation, which called itself the Australian Nationalist Workers Union, drawing on Eureka history (Holland 2003; Lentini 2019).

In its pursuit of ethnically homogeneous substate communities, the ANM was notorious for its violent tactics. However, similar ideas incorporating bioregionalist concepts of land and place connection were embraced by some extreme-right groups in Australia that did not engage in large public displays of violence or have substantial memberships. For example, Paul Innes, the former moderator of a Stormfront Downunder forum who went by the name 'Steelcap Boot' on Stormfront, imitated his brother David Innes's earlier attempt to establish a whites-only community in the Perth foothills as part of the Australian neo-Nazi Pioneer Little Europe (PLE) movement (Fleming 2009; Taylor 2009). Another group with comparable ideas was a collective of Tasmanian neo-Nazi biker metalheads who called themselves '*Waldkampf*', which loosely translates into a combination of 'fight' and 'forest'. Their propaganda also broadly references the neo-Nazi concept of promoting physical fitness in preparation for an anticipated race war and is influenced by the Nazi concept of *Volksgemeinschaft* or a 'people's community' (see Waldkampf 2022). *Waldkampf* members showcased their far-right beliefs through National Socialist Black Metal music and displaying esoteric Nazism symbols such as the *Sonnenrad*, or Sunwheel, on social media platforms like Facebook and Instagram. Such autonomous groups in regional Australian outposts might also be said to derive inspiration from international organisations with regionalist orientations, such as pan-Nordic neo-Nazi collective the Nordic Resistance Movement (Bjørgo and Ravndal 2020).

Similar, often superficial adaptations of bioregionalist notions of land and place connection were also present in an emerging NA movement across parts of the US, the UK, Australia and Europe, as well as elsewhere (Macklin 2005).[12] National anarchism can be understood as a political ideology that places a high value on decentralised, local communities, and it is anarchist only in name, rejecting the egalitarianism and anti-hierarchical positioning embraced historically by political anarchists. While also rejecting modern liberal capitalism and US hegemony, NAs tend to express solidarity with various socialist projects in the 'Third World' (Sunshine 2008). The movement's opposition to US political hegemony also aligns it with the further-right rejection of empires, as seen in the works of Francis Parker Yockey (Macklin 2005) and Jean-François Thiriart and their attempted alliance with both the Soviet Union and communist revolutionary movements in Asia and Latin America (Dugin 2014; Sunshine 2008). Antisemitism remains a fundamental aspect of NA groups, and this is sometimes reflected in expressions of solidarity with Arab anti-Zionists (Sunshine 2008).

For the purpose of this investigation, it is also necessary to briefly expound upon NA's tribalistic philosophy. Sharing similarities with ecofascists such as Pentti Linkola and Savitri Devi, its supporters nevertheless seek to distinguish their ideology from traditional fascism by incorporating critiques of capitalism and sometimes imitating the sustainable farming practices of marginalised Nazis Otto Strasser and Richard Walther Darré (Moore and Roberts 2022). Proponents of national anarchism also tend to draw inspiration from countercultural figures such as Ted Kaczynski, who was known as the 'Unabomber', and his anti-technology manifesto (Taylor 2019). NA organisations seek to establish a federalist model of governance that allows for greater autonomy at the local level, combined with an anti-civilisational, masculinist pagan ideology that emphasises traditional gender roles and the rejection of modern society (Griffin 2003). NA groups will present themselves as 'neither right nor left', adopting a Third Positionist approach that rejects both capitalist and communist systems, and in the manner of fascism proper, claim to advocate for a third alternative based on traditional values (Ross 2017, 103). The movement and its predecessors have been influenced by the European New Right (ENR) and National Bolshevism, a political ideology advocated by figures such as Aleksandr Dugin and Jean Parvulesco in Russia. NAs also draw inspiration from the 'Primordial Tradition' developed by the esoteric Italian fascist Julius Evola, which stresses the importance of returning to traditional and premodern societies; Ernst Jünger's 'revolutionary conservatism'; as well Friedrich Nietzsche's 'aristocratic radicalism' (Griffin 2003, 34).

To counter their limited presence in a political environment dominated by regionalist and internationalist white supremacist organisations after WWII, NA groups have attempted to cross-pollinate their ideologies with like-minded organisations and individuals around the world (Macklin 2005, 320). In particular, the NA movement has been known to exploit the anti-globalisation and anti-capitalist sentiments that arose in the anti-globalisation movements of the 1990s by using environmental issues as a guise for their far-right agenda against the alleged 'New World Order'. In the US, NA followers infiltrated left-leaning demonstrations and disrupted the efforts of organisations such as the Hunt Saboteurs Association and the Animal Liberation Front (Macklin 2005; Sunshine 2008). They have also participated in anti-imperial, pro-LGBTQIA+ and anti-globalisation protests. In Australia, for example, NAs protested against the Asia-Pacific Economic Cooperation (APEC) summit in September 2007 and for Tibet against the Chinese Government during the Olympic torch relay in April 2008 (Sunshine 2008), as well as during a Free Palestine rally at Sydney Town Hall in 2012. NA groups globally have correspondingly drawn influence from anarcho-pluralists such as Keith Preston in the US, who advocated for an alliance of left and right libertarians opposed to US imperialism and the state (Lyons 2011).[13] Richard Hunt, an anarcho-primitivist, was another figure formerly associated with the left who inspired prominent NAs – such as Troy Southgate from the UK National Revolutionary Faction (NRF). Hunt published a journal influential to national anarchism called *Green Alternative*, after he was expelled from the British Magazine *Green Anarchist* for endorsing the UK's involvement in the Persian Gulf War (see also Griffin 2005; Ross 2017, 103).[14]

NA and New Right groups in Australia more generally drew significant inspiration from the UK NRF. This was an influential group established in 1996 by Troy Southgate that originated from the former National Front, and the later English Nationalist Movement (Macklin 2005). Despite their exploitation of the international anti-globalisation movement, some NA groups, including the NRF, rejected the notion of building a mass social movement to disrupt global capitalism and imperialism as naive. Instead, they championed the idea of 'leaderless resistance' (Macklin 2005, 319).[15] While Third Positionism, decentralisation and regionalism characterised the NRF, the organisation also comprised a syncretic synthesis of tribalism, the notion of preserving Indo-European 'ethnic heritage' as a redress to globalisation, and Jungian notions of the collective subconscious, connected to the 'primeval Aryan psyche' (Macklin 2005, 308). The NA dispositions of the NRF included its adherents' (mis)appropriation of anarchist imagery and the superficial adoption of some of the more sophisticated critiques of global capitalism by anarchist philosophers such as Pierre-Joseph Proudhon, Pyotr Kropotkin and Mikhail Bakunin (Macklin 2005, 309–310).

The NA movement in Australia drew significant inspiration from the NRF, primarily through the New Right Australia/New Zealand (NRANZ), a small, now-defunct political group. The NRANZ incorporated ENR and identitarian ideals, claiming to be influenced by 'Traditionalists, Revolutionary Conservatives; the Nouvelle Droit [*sic*], and the Eurasianists', while rejecting liberalism, democracy and egalitarianism (New Right Australia/New Zealand 2013). One of the key figures in NRANZ was Welf Herfurth, a self-proclaimed NA who was involved in several far-right nationalist movements in his lifetime, acted as the master of ceremonies at an Australia First Party gathering in 2009 and was regarded as Australia's version of Troy Southgate. In 2007, NRANZ members disguised in black bloc protested at an anti-APEC summit (Fleming 2007). A propaganda video hosted on their website features members dressed in Guy Fawkes masks and black hoodies protesting outside Sydney Town Hall at a Free Palestine rally in 2012. Riffing on the trope of the left-wing Australian Trotskyist organisation, Socialist Alternative, the video presents NA as 'Beyond Left and Right. Above All Dogmas. For a Real Social Alternative. For The Community – Against The State. Think Outside The Box. National Anarchist Movement. SMASH ALL POLITICAL DOGMAS'. On the website, there is also an appreciative review by Keith Preston of Herfurth's

A Life in the Political Wilderness, a biographical memoir that outlines his perspectives on politics and activism.

Welf Herfurth's political activity in Australia provides an example of how national anarchism and similar ideologies could gain influence among authoritarian and ethnonationalist environmentalists. Herfurth, who was born in Germany and lived in Iran during the Islamic Revolution of 1979, joined the National Democratic Party upon returning to Germany and later, after moving to Australia, became involved with various other movements and organisations, including the Australian Democrats, ONP, the Australia First Party, Blood & Honour, and the Inverell and Sydney Forums (Fleming 2007; Beirich 2013). He was also a board member of the neo-Nazi skinhead group Blood & Honour Europe and a member of the violent, segregationist white supremacist organisation concentrated in America's Pacific Northwest, Volksfront. In 2013, the Southern Poverty Law Centre reported that Herfurth had a high-paying job with a shipping firm and was known for funding Volksfront projects and loaning individual members money. Herfurth's aforementioned book was well received in 'relatively highbrow racist circles like American Renaissance magazine', according to the Southern Poverty Law Center (Beirich 2013).

Although the NA movement failed to gain significant traction in Australia and the UK, later chapters in this book demonstrate how the environmentalism and racial separatism characteristic of the NAs have had a notable influence on prominent far-right and neo-fascist figures in Australia today, particularly in the context of climate change (see Chapter 3, p. 80). It is also relevant to note that the NA movement in Australia garnered momentum through blogs, a limited social media presence and public protests. NA has primarily spread through the internet, including on websites such as Stormfront, as part of an international white supremacist backlash against economic and cultural globalisation (Sanchez 2009). It is in this light associated with a version of palingenetic ultranationalism (Sunshine 2008), but also with the concept of 'folk autonomy' replacing traditional nationalism (Macklin 2005). Despite attempting to enter left-wing ecological movements and appropriating the ideas of anarchist philosophers and anarchist anti-capitalist principles and aesthetics, NA has been rejected by progressive anarchists and has thus far had a marginal influence among far- and extreme-right organisations.[16]

Concluding comments

This chapter has explored engagements with the natural environment and environmentalist politics on the part of far-right actors and movements in Australian political history. The discussion primarily focused on the events leading up to and following the 1901 Federation and the introduction of the Immigration Restriction Act, which, in its various forms and until its formal dissolution in 1973 by the ALP Whitlam government, prescribed almost exclusively European immigration to Australia (Jayasuriya 2012). The chapter illustrated that colonial genocide against Indigenous people played a critical role in shaping the Australian nationalist project (Langton 2020), which was further reinforced by the ongoing capitalist exploitation of Australian lands and territories, including through the relatively unconstrained extractive processes of mining and industrial agriculture (Walters 2023). The development of a mythic Australianist nativist and masculinist identity predicated on larrikinism, bushman ideologies and independence from Empire also informed the development of a republican white nationalism, in which the archetypal Australian figure was characterised by their spiritual and physical domination of the natural environment (Ward 1958).

The chapter analysed key historical events that have had a lasting impact on far-right environmentalist political thought in Australia, emphasising the periods of emergence for different far-right environmentalist, territorial and pseudo-environmentalist tendencies. The analysis first explored the mid-nineteenth century support for the exclusive rights of white workers within the emerging Australian labour movement, and then the union movement through the twentieth century (Burgmann 1984). It also considered the far-right reaction to worker movements by industrialists who held strategic interests in suppressing workers' bargaining power and undermining egalitarian institutions. This included on the part of fascist organisations such as the New Guard that lobbied in parliamentary arenas against NSW Labor Premier Jack Lang and enacted paramilitary violence against labour movement organisers (Cathcart 1988). These tendencies also fuelled a Cold War-era animosity towards workers' movements deemed 'socialist' or 'communist', occurring both in Australia and across the Western world (Smith 2020).

Entryism into left-wing political movements and their disruption was evidenced by various cases, including the ongoing early contestations between republican-labourist and imperial-industrialist forces in Australia. Syncretism, appropriation and disruption also existed outside industrial agitating. Cultural and literary organisations aligned with the interwar AFM aggressively and syncretically adapted Indigenous terminology and cultural references into their mission of developing a uniquely Australian nativist identity, which was also antisemitic and incorporated Aryanist and Odinist 'blood and soil' references (Bird 2014).

Throughout the mid to late twentieth century, ruralist movements like the ALR capitalised on anti-urban sentiments. These attitudes were historically significant for far-right and proto-fascist groups, and they gained renewed relevance in Australia after WWII due to far-right actors' reaction to industrialisation, expanding cities and increased non-Western European immigration (Davison and Brodie 2005; Collins 2008). The ALR strategically engaged in entryism, infiltrating the emerging Australian organic movement after WWII and making inroads into both the Liberal and Country (later National) Australian political parties (Gaynor 2012; Moore 2005). Alongside organisations such as the AAFI and NA, the ALR took advantage of anti-immigration politics that emerged alongside the modern Western environmentalist movement. Various dominant aspects of the movement attributed pollution, unsustainable practices and climate change to Global South population growth – though they only opposed unsustainable practices inasmuch as they go beyond the limits of sustainable capitalism (see e.g., Saleam 1999; Smith et al. 2023b).

The chapter also examined the significance of militarism and military histories for Australian far-right and white supremacist groups throughout history, highlighting the increasing importance of recruitment from RSLs for white nationalist groups from the 1980s (Kulmar and Jensen 2021). We also explored the tendencies and expressions of bioregionalist philosophy among NAs and other independent groups in the 2000s, following the propagandised promotion of autonomous whites-only communities by the ANM in the 1980s (Sunshine 2008; Fox 2023). While the appropriation of bioregionalist philosophy by the Australian far right may be superficial, it demonstrates another point of entryism and the opportunity for recruitment from the left by dedicated white supremacist organisations and individuals (Macklin 2005). The analysis speculated that this tendency could gain greater importance within far-right milieux as new, autonomous and substate responses to global heating emerge among communities that might not traditionally associate with the ethnonationalism and authoritarianism of the far right.

Throughout diverse historical periods, the Australian political landscape has featured the use of various far-right quasi-environmental themes, including white supremacist labourism, as well as a nativist rejection of, and ideological definition by, the country's status as a British colonial outpost. Subjective financial incentives for different far-right actors have also played a crucial role in shaping their approach to the natural environment and related treatment of quasi-environmentalist political ideas. Building on this exploration, the following chapter investigates political engagements with the environment and climate change in the propagandised online communications of contemporary far-right Australia-based actors. Drawing on examples from various cross-national websites and social media, this investigation examines the inflection of contemporary 'environmental values' (Bird 2014) in interactions between peripheral far-right actors and more dedicated, organised extreme-right groups.

Notes

1 In October 2018, Hanson also tabled a motion stating that 'It's OK to be white', titled after a white supremacist meme that first appeared on 8Chan, which was only defeated in the Senate 28 votes to 31. In June 2021, her motion to 'ban Critical Race Theory from a National Curriculum' passed 28 to 30 votes after the word 'ban' was replaced with 'restrict'.

2 The date Anning's staff selected for his speech was noted to feature the numbers 14 and 88, which in esoteric neo-Nazi symbolism are often taken to refer to the eighth letter of the alphabet repeated, denoting 'Heil Hitler', and David Lane's 14 words.

3 Third Positionism is a far-right political ideology that seeks to reject both capitalism and communism, instead advocating for a 'third way' that combines elements of both left-wing and right-wing politics. It emerged in the late twentieth century and has been associated with various far-right movements and organisations, primarily in Europe. Its core tenets variously include nationalism, an opposition to globalisation, authoritarianism, a rejection of liberal democracy, and economic self-sufficiency.

4 At least 11,074 Aboriginal and Torres Strait Islander people died directly from frontier massacres between 1788 and 1930 (Ryan et al. 2019).

5 The historical neglect of Indigenous knowledge and practices has had far-reaching consequences, not only for Indigenous peoples but also for the environment in Australia

as a whole. For example, traditional Indigenous fire management practices have been shown to prevent wildfires and reduce GHG emissions, yet these techniques have been suppressed and discouraged by colonial authorities for centuries (Kimmerer and Lake 2001). The lack of recognition and protection of Indigenous lands and waters has also led to many cases of environmental destruction, such as the degradation of the Murray-Darling Basin, a vital river system in south-eastern Australia (Beresford 2021).

6   Despite the prevalence of 'white nationalism' across the historical Australian political spectrum, Indigenous-led environmental and anti-colonial activism has challenged these narratives for a long time. In the current context, Indigenous activists and their allies are working to safeguard cultural heritage sites from industrial agriculture and mining while also advocating for the acknowledgement and preservation of Indigenous knowledge and environmental practices (Reynolds 2006).

7   Eureka is in some cases taken to represent an emblem of maritime navigation, by representation of the *Crux Australis* constellation of five stars in the Australian galaxy, overlain with a Christian cross. In other circumstances it was interpreted as inspired by the flags of the US and the UK, given that early miners in Australia were sometimes from these countries.

8   The CPA did call for immigration restrictions on people of certain nationalities.

9   Herbert Basedow, an Australian anthropologist, geologist and medical doctor, expounded the 'Black Caucasian' thesis in 1925. In his work, Basedow suggested that Indigenous Australians were not a separate racial group but rather a branch of the Caucasian race that had undergone physical adaptations due to environmental factors over time. Basedow believed that Indigenous Australians shared certain physical features and cultural practices with 'Caucasians', leading him to classify them under the same racial category. Basedow's ideas were based on certain racial theories of his time, which have since been discredited. Modern anthropologists and geneticists recognise that human populations cannot be neatly classified into distinct racial categories, and the concept of race itself is more of a social construct than a biological reality.

10 It is important to note that a lot of people were interned in Australia during WWII simply because of their nationality.

11 Strasserite refers to a political ideology associated with the brothers Otto and Gregor Strasser, who were prominent members of the Nazi Party in Germany during the 1920s and early 1930s. The Strasserite faction within the Nazi Party advocated for a more socialist version of national socialism, emphasising anti-capitalism and class struggle.

12 In Europe, some of the most well-known NA groups include the Autonomous Nationalists in Germany and the National Revolutionary Faction in the UK. In the US, some well-known NA groups include the Bay Area National Anarchists, based in California, and the National Anarchist Tribal Alliance, based in New York.

13 Preston's late ideological commitments were characterised by Lyons (2011) as an 'anti-humanist philosophy of elitism, ruthless struggle, and contempt for most people', which posed a dangerous alternative to other revolutionary forces.

14 Hunt's ideas, elaborated in his 1997 book *To End Poverty*, emphasised Wallerstein's theory that the 'periphery' or 'developing world' was exploited to benefit wealthy industrialised countries that make up the 'core'.

15 Since its founding, the NRF has increasingly incorporated occultist elements from the Nordic and Evolian countries and abandoned Catholicism, including its earlier advocacy of a 'wealth distribution system' known as 'Catholic distributivism' (Macklin 2005, 306).

16 Efforts at left-right coalition building by figures like Hunt and Southgate were met with opposition in the US and UK (Macklin 2005), while Herfurth and his co-agitators were publicly denounced by a collective of anarchists in Australia. On 21 October 2007, this collective issued a statement labelling self-proclaimed 'national anarchists' in Australia as 'imposters' (Fleming 2007). The statement criticised the use of black bloc tactics and the

appropriation of antifascist imagery, arguing that such tactics deliberately aimed to confuse the broader public about the aims and methods of far-right elements and to 'confuse anarchism with the pursuit of "white nationalist" politics'. The message further asserted that the adoption of such tactics aimed to draw people away from 'true anarchy; a society without rulers – of whatever color, and of whatever supposed nationality' (Fleming 2007).

References

Adamson WL (1980) 'Gramsci's interpretation of fascism', *Journal of the History of Ideas*, 41(4): 615–633.

Ashton P and Cornwall J (2006) 'Corralling conflict: the politics of Australian federal heritage legislation since the 1970s', *Public History Review*, 13: 53–65.

Atkinson A (1991) *Chinese labour and capital in Western Australia, 1847–1947* [PhD Thesis], Murdoch University.

Attwood B (2020) *The struggle for Aboriginal rights: a documentary history*, Routledge, London.

Australian Human Rights Commission (1997) *Bringing them home report*. https://humanrights.gov.au/our-work/bringing-them-home-report-1997

Australian Human Rights Commission (2021) *The racial hatred act: case study 3*. https://humanrights.gov.au/our-work/racial-hatred-act-case-study-3-0

Bedford L, McGillivray L and Walters R (2020) 'Ecologically unequal exchange, transnational mining, and resistance: a political ecology contribution to green criminology', *Critical Criminology*, 28: 481–499.

Beiner R (2018) *Dangerous minds: Nietzsche, Heidegger, and the return of the far right*, University of Pennsylvania Press, Pennsylvania.

Beirich H (20 November 2013) 'Volksfront: the leadership', *SPLC*, accessed 1 June 2022. https://www.splcenter.org/fighting-hate/intelligence-report/2013/volksfront-leadership-0

Bell A (2014) *Relating Indigenous and settler identities: beyond domination*, Springer, New York.

Beresford Q (2021) *Wounded country: the Murray-Darling basin – a contested history*, NewSouth Publishing, Randwick.

Bird D (2014) *Nazi dreamtime: Australian enthusiasts for Hitler's Germany*, Anthem Press, London.

Bjørgo T and Ravndal JA (2020) 'Why the nordic resistance movement restrains its use of violence', *Perspectives on Terrorism*, 14(6): 37–48.

Boukala S and Tountasaki E (2019) 'From black to green: analysing Le Front National's 'patriotic ecology'', in Smith E, Persian J and Fox VJ (eds) *The far right and the environment*, Routledge, London.

Burgmann V (1984) 'Racism, socialism, and the labour movement, 1887/1917', *Labour History*, 47: 39–54.

Burgmann V (1985) 'Who our enemies are: Andrew Markus and the baloney view of Australian racism', *Labour History*, 49: 97–101.

Burgmann V (2016) *Globalization and labour in the twenty-first century*, Taylor & Francis, London.

Busbridge R, Moffitt B and Thorburn J (2020) 'Cultural Marxism: far-right conspiracy theory in Australia's culture wars', *Social Identities*, 26(6): 722–738.

Butler E (30 November 1951) 'Developing an organic farm', *New Times*, 6(7): 253–269.

Butler E (January 1952) 'Developing an organic farm', *Victorian Compost News*, 6(1): 1–4

Butler E (1985) *The truth about the Australian league of rights*, Heritage Publications, Melbourne.

Campbell AA (1978) *The Australian league of rights: a study in political extremism and subversion*, Outback Press, Coolah.

Campbell E (1965) *The rallying point: my story of the new guard*, Melbourne University Press, Melbourne.

Campion K (2021) 'Defining ecofascism: historical foundations and contemporary interpretations in the extreme right', *Terrorism and Political Violence*, 35: 926–944. http://doi.org/10.1080/09546553.2021.1987895

Cathcart M (1988) *Defending the national tuckshop: the secret army intrigue of 1931*, McPhee Gribble/Penguin, Carlton.

Collins J (2008) 'Globalisation, immigration and the second long post-war boom in Australia', *The Journal of Australian Political Economy*, 61: 244–266.

Connolly CN (1978) 'Miners' rights', *Labour History*, 35: 35–47.

Considine M (1985) 'The national civic council: politics inside out', *Politics*, 20(1): 48–58.

Coupland PM (2016) *Farming, fascism and ecology: a life of Jorian Jenks*, Taylor & Francis, London.

Cronon W (1996) 'The trouble with wilderness: or, getting back to the wrong nature', *Environmental History*, 1(1): 7–28.

Crosby M (2005) *Power at work: rebuilding the Australian union movement*, Federation Press, Alexandria.

Cunneen C, Fraser D and Tomsen S (eds) (1997) *Faces of hate: hate crime in Australia*, Hawkins Press, Annandale, NSW.

Curthoys A (1999) 'Expulsion, exodus and exile in white Australian historical mythology', *Journal of Australian Studies*, 23(61): 1–19.

Curthoys A and Moore C (1995) 'Working for the white people: an historiographic essay on Aboriginal and Torres Strait Islander labour', *Labour History*, 69: 1–29.

Dalley C (2022) 'Pastoralism's distributive ruse: extractivism, financialization, Indigenous labour and a rightful share in Northern Australia', *History and Anthropology*. https://doi.org/10.1080/02757206.2022.2034622

Davison G and Brodie M (ed) (2005) *Struggle country: the rural ideal in twentieth century Australia*, Monash University ePress, Melbourne.

Dugin A (2014) *Eurasian mission: an introduction to neo-Eurasianism*, Arktos, Budapest.

Elder C (2020) *Being Australian: narratives of national identity*, Routledge, London.

Elias A, Mansouri F and Paradies Y (2021) *Racism in Australia today*, Palgrave Macmillan, London.

Encel S (1989) 'Antisemitism and prejudice in Australia', *Patterns of Prejudice*, 23(1): 16–27.

Fleming A (2007) 'Anarchist statement on the new right', *Slackbastard*. https://slackbastard.anarchobase.com/?p=908

Fleming A (2009) 'Police monitor supremacist group in Perth Hills', *Slackbastard*. https://slackbastard.anarchobase.com/?p=1682

Fleming A and Mondon A (2018) 'The radical right in Australia', in Rydgren J (ed) *The Oxford handbook of the radical right*, Oxford University Press, Oxford.

Fleming G, Merrett D and Ville S (2004) *The big end of town: big business and corporate leadership in twentieth-century Australia*, Cambridge University Press, Cambridge.

Forchtner B and Kølvraa C (2015) 'The nature of nationalism: Populist radical right parties on countryside and climate', *Nature and Culture*, 10(2): 199–224.

Fox VJ (2022) 'Far-right terror: Perth's 1980s 'fascist revolution'', *Red Flag*, accessed 1 June 2022. https://redflag.org.au/article/far-right-terror-perths-1980s-fascist -revolution

Fox VJ (2023) 'Fascism and anti-fascism in Perth in the 1980s', in Smith E, Persian J and Fox V (eds) *Histories of fascism and anti-fascism in Australia*, Routledge, London.

Funston L and Herring S (2016) 'When will the stolen generations end?: a qualitative critical exploration of contemporary 'child protection' practices in Aboriginal and Torres Strait Islander communities', *Sexual Abuse in Australia and New Zealand*, 7(1): 51–58.

Garton S (2010) 'Eugenics in Australia and New Zealand: laboratories of racial science', in Bashford A and Levine P (eds) *The Oxford handbook of the history of eugenics*, Oxford University Press, Oxford.

Gaynor A (2012) 'Antipodean eco-nazis? The organic gardening and farming movement and far-right ecology in postwar Australia', *Australian Historical Studies*, 43(2): 253–269.

Gillespie L (2020) 'The imagined immunities of defense nationalism', *Political Psychology*, 41(5): 997–1011.

Gordon SA (1984) *Hitler, Germans, and the 'Jewish question'*, Princeton University Press, New Jersey.

Graham BD (1966) *The formation of the Australian country parties*, Australian National University Press, Canberra.

Greason D (1997) 'Australia's racist far-right', in Cunneen C, Fraser D and Tomsen S (eds) *Faces of hate: hate crime in Australia*, Hawkins Press, Sydney.

Grégoire ER and Hatcher P (2022) 'Global extractivism and inequality', in Sims K, Banks N, Engel S, Hodge P, Makuwira, J, Nakamura N, Rigg J, Salamanca A and Yeophantong P (eds) *The Routledge handbook of global development*, Routledge, London.

Griffin N (2005) 'Nick Griffin, "national-anarchism: Trojan horse for white Nationalism"', *Green Anarchy*, 19, accessed 1 June 2023. http://web.archive.org/web/20090825115331 /http://www.greenanarchy.org:80/index.php?action=viewwritingdetail&writingId=150

Griffin R (2003) 'From slime mould to rhizome: an introduction to the groupuscular right', *Patterns of Prejudice*, 37(1): 27–50.

Griffiths B (2018) *Deep time dreaming: uncovering ancient Australia*, Black Inc, Melbourne.

Griffiths T (2016) *The art of time travel: historians and their craft*, Black Inc, Melbourne.

Hage G (2012) *White nation: fantasies of white supremacy in a multicultural society*, Routledge, London.

Hage G (2017) *Is racism an environmental threat?* Polity Press, Cambridge.

Hanebrink P (2018) *A specter haunting Europe: the myth of Judeo-Bolshevism*, Harvard University Press, Cambridge.

Hanson PL and Merritt GJ (1997) *Pauline Hanson – the truth: on Asian immigration, the Aboriginal question, the gun debate and the future of Australia*, St George Publications, Parkholme.

Hartley L and Fleay C (2017) ''We are like animals': negotiating dehumanising experiences of asylum-seeker policies in the Australian community', *Refugee Survey Quarterly*, 36(4): 45–63.

Harvey D (1996) *Justice, nature and the geography of difference*, Blackwell, London.

Harvey D (2005) *The new imperialism*, Oxford University Press, Oxford.

Head L (2022) 'Beyond green: human–environment geographies for the 'new' century', *Environment and Planning F*, 1(1): 93–103.

Henderson G (2015) *Santamaria: a most unusual man*, Melbourne University Publishing, Melbourne.

Henderson P (2005) 'Frank Browne and the neo-Nazis', *Labour History*, 89: 73–86.

Hickel J, O'Neill DW, Fanning AL and Zoomkawala H (2022) 'National responsibility for ecological breakdown: a fair-shares assessment of resource use, 1970–2017', *The Lancet Planetary Health*, 6(4): e342–e349.

Holland I (2003) *Crime and candidacy.* https://apo.org.au/sites/default/files/resource-files /2003-03/apo-nid6715.pdf

Hu G (2022) 'Trauma writing in the gold fields by Ouyang Yu', *Open Access Library Journal*, 9(9): 1–5.

Hughes B, Jones D and Amarasingam A (2022) 'Ecofascism: an examination of the far-right/ecology nexus in the online space', *Terrorism and Political Violence*, 34(5): 997–1023.

Human Rights Watch (2021) *Australia: act on Indigenous deaths in custody*, accessed 1 June 2022. https://www.hrw.org/news/2021/04/14/australia-act-indigenous-deaths-custody

Hunt D (1978) 'Exclusivism and unionism', *Labour History*, 35: 80–95.

James S (2005) 'The policing of right-wing violence in Australia', *Police Practice and Research*, 6(2): 103–119.

Jayasuriya L (2012) *Transforming a 'White Australia': issues of racism and immigration*, SSS Publications, New Delhi.

Johns A (2008) 'White tribe: echoes of the Anzac myth in Cronulla', *Continuum*, 22(1): 3–16.

Jones C (2021) 'Conspiracy theories and the Australian far right', *Monash Lens*, accessed 1 June 2022. https://lens.monash.edu/@politics-society/2021/02/04/1382797/conspiracy -theories-and-the-australian-far-right

Jones R (2008) 'Soil: a real and imagined environment for Australian organic farmers and gardeners in the 1940s', *Environment and History*, 14(2): 205–215.

Jones R and Gaynor A (2019) 'Wild harvesting, self-sown crops, and the ambiguous modernity of Australian agriculture', *Agricultural History*, 93(2): 212–232.

Jupp J (2002) *From white Australia to Woomera: the story of Australian immigration*, Cambridge University Press, Cambridge.

Keane B (9 November 2022) 'The liberal party faces two paths: moderate liberalism or republican extremism', *Crikey*, accessed 1 June 2022. https://www.crikey.com.au/2022 /11/09/liberal-party-future-republican-extremism-or-moderate-liberalism/

Kiernan B (2008) *Blood and soil: a world history of genocide and extermination from Sparta to Darfur*, Yale University Press, Connecticut.

Kimmerer RW and Lake FK (2001) 'The role of Indigenous burning in land management', *Journal of Forestry*, 99(11): 36–41.

Kulmar C and Jensen M (24 January 2021) 'Far right and extremist groups are targeting military veterans for recruitment: does the ADF owe them a duty of care?', *The Conversation*, accessed 1 June 2022. https://researchsystem.canberra.edu.au/ws/files /53816573/Far_right_and_extremist_groups_are_targeting_military_veterans.pdf

Kwok J (2022) 'The lambing flat riots and the Chinese quest for compensation', *Journal of Australasian Mining History*, 20: 86–107.

Laney HM, Lenette C, Kellett AN, Smedley C and Karan P (2016) '"The most brutal immigration regime in the developed world": international media responses to Australia's asylum-seeker policy', *Refuge*, 32: 135.

Lake M and Reynolds H (2010) *What's wrong with Anzac? The militarisation of Australian history*, University of NSW Press, Sydney.

Langton M (2020) 'Welcome to country: knowledge', *Agora*, 55(1): 3–10.

LeCras L (2023) '"War on the whiteman": Australia in the racial and political imagination of Britain's far right, 1945–1970', in Smith E, Persian J and Fox VJ (eds) *Histories of fascism and anti-fascism in Australia*, Routledge, London.

Leibler IJ (1967) 'John Birch down under', *Patterns of Prejudice*, 1(2): 25–26.

Lentini P (2019) 'The Australian far-right: an international comparison of fringe and conventional politics', in Peucker M and Smith D (eds), *The far-right in contemporary Australia*, Palgrave Macmillan, Singapore.

Lindqvist S (2007) *Terra nullius: a journey through no one's land*, Granta, London.

Lipscombe TA, Dzidic PL and Garvey DC (2020) 'Coloniser control and the art of disremembering a "dark history": duality in Australia Day and Australian history', *Journal of Community & Applied Social Psychology*, 30(3): 322–335.

Lowe D (1999) *Menzies and the 'great world struggle': Australia's Cold War, 1948–1954*, UNSW Press, Sydney.

Lubarda B (2020) 'Beyond ecofascism? Far-right ecologism (FRE) as a framework for future inquiries', *Environmental Values*, 29(6): 713–732.

Lyons MN (2011) 'Rising above the herd: Keith Preston's authoritarian anti-statism', *New Politics*, 7(3).

MacDonald K (22 January 2010) 'Eureka flag "hijacked by racists"', *The West Australian*, accessed 1 June 2022. https://thewest.com.au/news/wa/eureka-flag-hijacked-by-racists-ng-ya-224116

Macintyre S (2004) *A concise history of Australia*, Cambridge University Press, Cambridge.

Macintyre S and Clark A (2013) *The history wars*, Melbourne University Publishing, Melbourne.

Macklin G (2005) 'Co-opting the counter culture: Troy Southgate and the National Revolutionary Faction', *Patterns of Prejudice*, 39(3): 301–326.

Macklin G (2020) *Failed Führers: a history of Britain's extreme right*, Routledge, London.

Markey R (1978) 'Populist politics', *Labour History*, 35: 66–79.

Markus A (1985) 'Explaining the treatment of non-European immigrants in nineteenth century Australia: comment', *Labour History*, 48: 86–91.

Markus A (13 April 2018) 'The far-right's creeping influence on Australian politics', *The Conversation*, accessed 1 June 2022. https://theconversation.com/the-far-rights-creeping-influence-on-australian-politics-93723

Martínez J (1999) 'Questioning "white Australia": unionism and "coloured" labour, 1911–37', *Labour History*. https://doi.org/10.2307/27516625

McFadden A (2023) 'Wardens of civilisation: the political ecology of Australian far-right civilisationism', *Antipode*, 55(2): 548–573.

McKenna M and Ward S (2007) '"It was really moving, mate": the Gallipoli pilgrimage and sentimental nationalism in Australia', *Australian Historical Studies*, 38(129): 141–151.

McKnight D and McChesney RW (2013) *Murdoch's politics: how one man's thirst for wealth and power shapes our world*, Pluto Press, London.

McNamara N (2014) 'Australian Aboriginal land management: constraints or opportunities', *James Cook University Law Review*, 21: 25.

Mendes P (2022) 'Combatting anti-semitism on the far right: an examination of the Jewish council to combat fascism and anti-semitism and its anti-defamation activities', *Australian Jewish Historical Society Journal*, 26(1): 54–78.

Merrill MC (1983) 'Eco-agriculture: a review of its history and philosophy', *Biological Agriculture & Horticulture*, 1(3): 181–210.

Metcalfe L (2013) 'The impact of "White Australia" on the development of Australian national identity in the period between 1880 and 1914', *History Initiates*, 1(1): 2–9.

Miller RJ, Ruru J, Behrendt L and Lindberg T (2010) *Discovering Indigenous lands: the doctrine of discovery in the English colonies*, Oxford University Press, Oxford.

Mitropoulos A (7 February 2018) 'Eureka: the "White Australia" policy and the holy relic of militant white liberalism', *sometim3s*, accessed 1 June 2022. https://s0metim3s.com /2018/02/07/eureka-flag/

Moore A (1989) *The secret army and the premier: conservative paramilitary organisations in New South Wales, 1930–32*, UNSW Press, Sydney.

Moore A (2005) 'Writing about the extreme right in Australia', *Labour History*, 89: 1–15.

Moore JW (ed) (2016) *Anthropocene or capitalocene?: nature, history, and the crisis of capitalism*, PM Press, Oakland.

Moore S and Roberts A (2022) *The rise of ecofascism: climate change and the far right*, Polity Press, Cambridge.

Moses AD (ed) (2004) *Genocide and settler society: frontier violence and stolen Indigenous children in Australian history*, Berghahn Books, New York.

Munro C (1984) *Wild man of letters: the story of PR Stephensen*, Melbourne University Press, Melbourne.

Murrie L (1998) 'The Australian legend: writing Australian masculinity/writing "Australian" masculine', *Journal of Australian Studies*, 22(56): 68–77.

Nairn B (1969) 'The 1916–17 labor party crisis in New South Wales and the advent of WJ McKell', *Labour History*, 16: 3–13.

National Inquiry into Racist Violence (1991) *Report of the national inquiry into racist violence in Australia*, Australian Government Publishing Service, Canberra.

New Right Australia/New Zealand (2013) *New right Australia/New Zealand*, accessed 1 June 2022. https://newrightausnz.blogspot.com/

Nilan P (2019) 'Far-right contestation in Australia: soldiers of Odin and True Blue Crew', in Peucker M and Smith D (eds) *The far-right in contemporary Australia*, Springer, New York.

Obaidi M, Kunst J, Ozer S and Kimel S Y (2022) 'The "great replacement" conspiracy: how the perceived ousting of whites can evoke violent extremism and Islamophobia', *Group Processes & Intergroup Relations*, 25(7): 1675–1695.

O'Carroll J (2020) *Borderwork in multicultural Australia*, Routledge, London.

Ong A (2006) *Neoliberalism as exception: mutations in citizenship and sovereignty*, Duke University Press, North Carolina.

Paull J (2013) 'A history of the organic agriculture movement in Australia', in Mascitelli B and Lobo A (eds) *Organics in the global food chain*, Connor Court Publishing, Ballarat.

Pearson W and O'Neill G (2009) 'Australia day: a day for all Australians?', in McCrone D and McPherson G (eds) *National days: constructing and mobilising national identity*, Palgrave Macmillan, London.

Peucker M, Lentini P, Smith D and Iqbal M (2021) '"Our diggers would turn in their graves": nostalgia and civil religion in Australia's far-right', *Australian Journal of Political Science*, 56(2): 189–205.

Piccini J and Smith E (2019) '"The "White Australia" policy must go": the Communist Party of Australia and immigration restriction', in Piccini J, Smith E and Worley M (eds) *The far left in Australia since 1945*, Routledge, London.

Piccini J, Smith E and Worley M (eds) (2018) *The far left in Australia since 1945*, Routledge, London.

Pullen JM and Smith GO (1997) 'Major Douglas and social credit: a reappraisal', *History of Political Economy*, 29(2): 219.

Rademaker L (2022) '60,000 years is not forever: "time revolutions" and Indigenous pasts', *Postcolonial Studies*, 25(4): 545–563.

Reynolds H (2006) *The other side of the frontier: Aboriginal resistance to the European invasion of Australia*, UNSW Press, Sydney.

Richards I (2019) 'A dialectical approach to online propaganda: Australia's United Patriots Front, right-wing politics, and Islamic State', *Studies in Conflict & Terrorism*, 42(1–2): 43–69.

Richards I and Jones C (2023) 'Far-right identitarianism in Australia', in Zúquete JP and Marchi R (eds), *Global Identitarianism*, Routledge, London.

Richmond K (1977) 'The Australian League of Rights and the process of political conversion', *Politics*, 12(1): 70–77.

Rickard J (2017) *Australia: a cultural history*, Monash University Publishing, Melbourne.

Rimmer SH (3 May 2022) 'Clive Palmer, his money and his billboards are back: what does this mean for the 2022 federal election?', *The Conversation*, accessed 1 June 2022. https://theconversation.com/clive-palmer-his-money-and-his-billboards-are-back-what -does-this-mean-for-the-2022-federal-election-182123

Rose DB (2011) *Wild dog dreaming: love and extinction*, University of Virginia Press, Charlottesville.

Rose DB and Australian Heritage Commission (1996) *Nourishing terrains: Australian Aboriginal views of landscape and wilderness*, Australian Heritage Commission, Canberra.

Ross AR (2017) *Against the fascist creep*, AK Press, Edinburgh.

Rudner J (2019) 'Hijacking democracy? Spatialised persecution and the planning process', in Peucker M and Smith D (eds) *The far-right in contemporary Australia*, Routledge, London.

Ryan L, Debenham J, Pascoe B, Smith R, Owen C, Richards J, Gilbert S, Anders R J, Usher K, Price D, Newley J, Brown M, Le L H, and Fairbairn H (2019) *Colonial frontier massacres in Australia 1788–1930*, University of Newcastle, Newcastle, accessed 1 January 2023. http://hdl.handle.net/1959.13/1340762

Sale K (2000) *Dwellers in the land: the bioregional vision*, University of Georgia Press, Georgia.

Saleam J (1999) *The other radicalism: an inquiry into contemporary Australian extreme right ideology, politics and organisation 1975–1995* [PhD Thesis], University of Sydney. https://ses.library.usyd.edu.au/bitstream/handle/2123/807/adt-NU20020222 .14582202whole.pdf;jsessionid=B8E56F208AD9A431F7820B38A025AD71 ?sequence=1

Sanchez C (29 May 2009) 'California racists claim they're anarchists', *SPLC*, accessed 1 June 2022. https://www.splcenter.org/fighting-hate/intelligence-report/2009/california -racists-claim-they're-anarchists

SBS News (15 August 2018) 'Full text: Senator Fraser Anning's maiden speech', *SBS News*, accessed 1 June 2022. https://www.sbs.com.au/news/full-text-senator-fraser-anning-s -maiden-speech/ba0d8a0e-702b-4bed-971e-7f9556a13aea

Seccombe M (10–16 November 2018) 'Neo-nazi Nats and party infiltrations', *The Saturday Paper*, accessed 1 June 2022. https://www.thesaturdaypaper.com.au/news/politics/2018 /11/10/neo-nazi-nats-and-party-infiltrations/15417684007114#hrd

Sewell B (1995) *'Australia full: Asians out! White supremacists in!': a study of the dynamics of the Australian National Action Movement in Australia* [PhD Thesis], Victoria University of Technology.

Shekhovtsov A (2017) *Russia and the western far right: Tango noir*, Routledge, London.

Silverstein J (2023) *Cruel care: a history of children at our borders*, Monash University Press, Melbourne.

Slater L (2018) *Anxieties of belonging in settler colonialism: Australia, race and place*, Routledge, London.

Smith C and Iner D (2021) 'How the contemporary far-right have popularised their appeals: an analysis of far-right growth in the Australian context', *Australian Journal of Islamic Studies*, 6(2): 1–30.

Smith E (2020) 'White Australia alone?: the international links of the Australian far right in the Cold War era', in Geary D, Sutton J and Schofield C (eds) *Global white nationalism: from apartheid to Trump*, Manchester University Press, Manchester.

Smith E, Persian J and Fox VJ (2023a) 'Introduction: fascism and anti-fascism in Australian history', in Smith E, Persian J and Fox VJ (eds) *Histories of fascism and anti-fascism in Australia*, Routledge, London.

Smith E, Persian J and Fox VJ (eds) (2023b) *Histories of fascism and anti-fascism in Australia*, Routledge, London.

Sparrow J (22 July 2015) 'If you oppose Reclaim Australia, remember fascism wasn't always a freakshow', *The Guardian*, accessed 1 June 2022. https://www.theguardian .com/commentisfree/2015/jul/22/if-you-oppose-reclaim-australia-remember-fascism -wasnt-always-a-freakshow?CMP=Share_iOSApp_Other

Sparrow J (15 August 2022a) 'Against populationism', *Overland*, accessed 1 June 2022. https://overland.org.au/2022/08/against-populationism/

Sparrow J (2022b) '"That's what drives us to fight": labour, wilderness and the environment in Australia', *Overland*, accessed 1 January 2023. https://overland.org.au/previous-issues /issue-246/feature-thats-what-drives-us-to-fight-labour-wilderness-and-the-environment -in-australia/

Stephensen PR (1935) *The foundations of culture in Australia*. https://nativistherald.com.au /2018/12/01/the-foundations-of-culture-in-australia/

Sunshine S (2008) 'Rebranding fascism: national-anarchists', *The Public Eye Magazine*, 23(4). https://files.libcom.org/files/Rebranding%20fascism.pdf

Sunter AB (2000) 'Something borrowed, something blue: the tale of a much-travelled piece of the Eureka flag, finally come home', *Overland*, 160: 69–71.

Tavan G (2012) 'Leadership: Arthur Calwell and the post-war immigration program', *Australian Journal of Politics & History*, 58(2): 203–220.

Taylor B (2019) 'Alt-right ecology: ecofascism and far-right environmentalism in the United States', in Forchtner B (ed) *The far right and the environment*, Routledge, London.

Taylor P (7 February 2009) 'Police monitor supremacist group in Perth Hills', *News.com.au*.

Teitelbaum B (2019) 'Daniel Friberg and metapolitics in action', in Sedgwick M (ed) *Key thinkers of the radical right: behind the new threat to liberal democracy*, Oxford University Press, Oxford.

Tranter B and Donoghue J (2007) 'Colonial and post-colonial aspects of Australian identity', *The British Journal of Sociology*, 58(2): 165–183.

Waldkampf (2022) https://www.instagram.com/waldkampf/

Waling A (2019) *White masculinity in contemporary Australia: the good ol' Aussie bloke*, Routledge, London.

Walters R (2023) 'Ecocide, climate criminals and the politics of bushfires', *The British Journal of Criminology*, 63(2): 283–303.

Ward R (1958) *The Australian legend*, Oxford University Press, Oxford.

Ward R (1978) 'Australian legend re-visited', *Australian Historical Studies*, 18(71): 171–190.

Wentworth WC (1931) 'Douglas social credit', *Australian Quarterly*, 3(12): 61–74.

WhiteRoseSociety (2019) *Fraser Anning's neo-Nazi connections*. https://thewhiterosesociety .writeas.com/fraser-annings-neo-nazi-connections

Whitford T (2011) 'A political history of national action: its fears, ideas, tactics and conflicts', *Rural Society*, 20(2): 216–226.

Whitford T (2013) 'Combating political police: an overview of National Action's counterintelligence program 1982–1990', *Salus Journal*, 1(2): 40–51.

Wilson E (1992) 'The invisible flâneur', *New Left Review*, 191(1): 90–110.

Winter B (2005) *The Australia first movement*, Interactive Publications, Carindale.

Zinn H (2015) *A people's history of the United States: 1492–present*, Routledge, London.

3 Ecofascism online

Australian far-right actors' use of environmental politics on cross-national media

Introduction

Ecofascism is often said to entail right-wing activists and ethnonationalist governments advocating severe population control measures (Biehl and Staudenmaier 2011) and accelerationist propaganda hastening the social and economic collapse of societies worldwide (Anker and Witoszek 1998). Although ecofascism and other far-right environmental ideologies are frequently examined as isolated phenomena, their core tenets arise from political–ideological contexts in which media and political actors in the Global North have erroneously attributed responsibility for climate change to the Global South through rhetoric about overpopulation and excess fossil fuel use (Moore and Roberts 2022). This has occurred despite wealthy countries historically driving planet-warming emissions through extractivist resource-intensive industries (Hickel et al. 2022), while many countries in the Global South lack the levels of industrial development necessary for sustainable living in the context of global heating (Blaustein et al. 2020).

The emergence of ecofascism and other far-right environmental ideologies have had particular significance for Australia, given its geopolitical situation, severe experience of global heating, restrictive immigration policies and politically motivated violence emanating from the country's political situation (Lockwood 2018; Sparrow 2019; McFadden 2022). The actions of Brenton Tarrant, an Australian self-proclaimed 'ecofascist', illustrated this in a high-profile case when he killed 51 people in a mass shooting at two mosques in Christchurch in 2019. Tarrant's motivation partly stemmed from the genocidal rhetoric of a 'global population cap' found in Anders Breivik's (2011) anti-Islam manifesto – Breivik being a neo-Nazi responsible for killing 77 people in 2011 through shootings and bombings in Utøya and Oslo, Norway. Tarrant's actions also subsequently inspired other attackers (Harwood 2021). One such individual was Patrick Crusius, who carried out a mass shooting at a Walmart store in El Paso, Texas, in August 2019, killing 23 people, mainly of Hispanic and Latino descent. Crusius titled his manifesto *The Inconvenient Truth* after Al Gore's climate change documentary (Richards and Wood 2020). In a later incident, in May 2022, 18-year-old Payton Gendron allegedly executed a mass shooting targeting black individuals at a Buffalo supermarket, resulting in 10 fatalities. Gendron's manifesto heavily

DOI: 10.4324/9781003325437-3

plagiarised Tarrant's manifesto, *The Great Replacement*, and he portrayed his attack as in the interests of 'green nationalism' (Richards et al. 2022).

Declarations of ecological-environmental concern by far-right actors have recently come to greater public prominence due to these actors' utilitarian fixation on climate change (Forchtner 2019a) and in light of a recent global surge in white supremacist mass violence (Pantucci and Ong 2021; GTD 2022). These actors' stated priorities are, however, often relatively uncritically accepted in media and policy circles, while the term 'ecofascism' has also increasingly featured in imprecise popular descriptions of right-wing governments' anti-immigration policies (Lubarda 2020; Thomas and Gosink 2021). There has been limited empirical research addressing the impacts of far-right political entryism in domains of environmental advocacy, or the mainstreaming of ecofascist ideas through government policy and popular political discourse (Sunshine 2008; Rueda 2020). An investigation into these effects, such as that set out in this book, is particularly relevant in Australia, where ethnocultural exclusivism and white supremacy remain prevalent, and pose ongoing challenges to institutional and extra-institutional responses to anticipated extreme future climate impacts (Price 2018b).

Building on the preceding chapter's examination of far-right actors' uses of environmental politics in Australian history, this chapter critically investigates current far- and extreme-right Australia-based actors' propagandistic engagement with environmental politics on cross-national online media platforms. The empirical analysis provides insights that inform a critical theoretical reflection on the relationship between ecofascism and other far-right political ideologies featuring nativist and neo-Nazi elements, which are common among Australian far- and extreme-right groups.

In this context, the discussion contemplates how the term 'ecofascism' has often been misused for political vilification, with its legitimacy generally dependent on the speaker's (1) emphasis on 'racial miscegenation' as a form of pollution (Byrne 2020), and (2) intent to establish authoritarian governance in response to ecological-environmental crises (Zimmerman 1995). The chapter acknowledges that traditional ecofascist discourse frequently incorporates blood and soil motifs, eugenics and radical antihumanism (Taylor 2019), while populationist policies implemented both within and outside governments are occasionally labelled ecofascist (Dyett and Thomas 2019).

The chapter begins from the premise that the pseudo-intellectual commitments of far-right actors, including the pseudo-environmentalist tendencies of ecofascism, operate in connection with broader political networks of climate change responses to legitimise various forms of far-right violence (Mudde 2019). This takes place through both state and non-state means, and in institutional and non-institutional political venues (e.g. Schwartz and Randall 2003). Through its contextual and comparative approach, the chapter critically analyses the facility far-right ecological tendencies including ecofascism provide for a wider far-right political entryism, which they also depend upon to flourish.

Consistent with its inductive orientation, this research does not assume that media or political actors primarily aim to legitimise far-right violence, to either a physical or symbolic extent. However, the analysis does investigate themes, especially within popular responses to climate change, that are vulnerable to co-optation by the far right (Hartmann 2010). The analysis also begins by considering that only a limited number of actors might be considered genuinely 'ecofascist' in nature, although currents of far-right and other political ideologies exploitative of climate change can, in some ways, be understood as aligning with ecofascist principles, if not always integrative of a holistic ecofascist vision (Forchtner et al. 2018).

Following this introductory section, there is a conceptual discussion of the various meanings of ecofascism and an examination of broader political environmentalist themes exploited by far- and extreme-right political actors, focusing on countries with European and settler-colonial histories. Next, a brief overview of extreme-right environmentalist expressions appearing in manifestos produced by high-profile white supremacist mass attackers is provided. This is followed by an analysis of the political-ideological media environments from which ecofascism and other violent tendencies emerge. The first part of this analysis examines Australia-based climate change discourse on the first dedicated online white nationalist website, Stormfront, focusing on discussions across three Australia-related forums: General Downunder Discussion, Downunder Newslinks, and Politics & Activism in Downunder.

The next part of the analysis examines how the administrators of the Australian Natives Association Facebook page and related pages, such as Australian Ecofascist Memes, engage with pseudo-environmentalist tropes. The 'Australia' group on the far-right social media site Gab is then explored, with a particular focus on commentary about climate change related to domestic political circumstances in Australia and regionally. In the final portion of the analysis, we examine the land- and environment-related activities of Australian extreme-right 'groupuscules' (Griffin 2003) active on Telegram,[1] focusing on members of the National Socialist Network (NSN) and its affiliate organisation, the European Australian Movement (EAM). The chapter concludes by providing succinct observations on the insights gleaned about ecofascism and other forms of far-right environmental politics exploited in the extra-institutional Australian context, as evidenced through far-right and extreme-right online media.

Conceptualising ecofascism

Ecofascism can best be understood as an extremist trope or sensibility rather than a coherent political ideology (Ross and Bevensee 2020; Hughes et al. 2022), and the meaning of the term tends to vary depending on its usage. Ecofascism now frequently refers to fascists and other far-right figures who incorporate superficial, insincere or opportunistic green components into their ideology (Biehl and Staudenmaier 2011; Dyett and Thomas 2019). However, the term has also been used to describe groups or individuals with a significant environmentalist aspect in their ideology, which is otherwise characterised by nativist, xenophobic (particularly anti-immigration) or even genocidal racism (Galbreath and Auers 2009). Historically, ecofascism has signified 'a radical blend of ethnonationalism and authoritarianism, rooted in a belief that *the land* and *the people* are symbiotically interwoven, and form an organic whole' (Forchtner 2019b). In the contemporary context, it often denotes far-right and extreme-right actors exploiting concern over the adverse effects of climate change (Allison 2020; Karbevski 2020).

Individuals and movements on the extreme right are occasionally labelled as ecofascist due to their taking inspiration from the environmental policies of the National Socialist German Workers' Party during WWII or, in more limited instances, the National Fascist Party in Italy (Atkins and Menga 2022). While modern government policies on immigration and development are also sometimes described as ecofascist, individuals referred to as such are generally non-state actors, although they frequently echo the principles of historical fascist statecraft, adjusted to the national context (Smith 2021).

Building on the previous chapter's discussion, the subsequent discourse and content analysis emphasises the importance of considering historical precedents when examining the use of environmental politics for propaganda purposes by far-right and extreme-right actors in Australia. It is crucial to acknowledge that environmentalism has often been secondary to other aspects of fascist doctrine. For instance, the so-called green wing of the Nazi Party was defeated by 1941, following Hitler's dissolution of the Strasser faction and *Sturmabteilung*[2] during the Night of the Long Knives in 1934. The environmental plans of the Reich Minister of Food and Agriculture, Richard Darré, were also consistently subordinated to the Nazi Party's militaristic exploitation of the natural environment and pursuit of capitalist expansion (Staudenmaier 2011a,b; Moore and Roberts 2022). While Italy's National Fascist Party may be perceived as dominators or 'conquerors' of the natural environment (Campion 2023), agricultural reclamation in Fascist Italy was intimately connected to the idea of national and cultural reclamation. Within this framework, land clearing by peasants was pursued for national economic ends, and ideologically it served as a model for the physical and spiritual regeneration of the Italian population along racial and national lines (Armiero 2014).

This chapter investigates the role of environmentalist politics in contemporary far- and extreme-right groups with diverse nativist, white nationalist and national socialist orientations in the Australian context, while analysing the broader ideological–political landscape from which ecofascism arises. The inquiry therefore implicitly examines the conceptual distinction between 'fascist first' and 'green first' ideologies linked to ecofascism and the ongoing theoretical debates concerning ecofascism's connections to deep ecologism (Bookchin 1987; Biehl and Staudenmaier 1995; Thomas and Gosink 2021).

Deep ecology[3] not only influenced US 'green' and 'brown' tribal movements during the 1980s (Sunshine 2008; Lyons 2011; Taylor 2019), but also inspired self-proclaimed NAs in Australia and the UK throughout the 1990s and 2000s (see Chapter 2, p. 51; Macklin 2005; Carter 2012; Schlembach 2013). However, deep ecologists often defend biocentrism against ecofascism (e.g. Orton 2000; Brown 2005). Although there is more overlap between ecofascism and deep ecology in the activities of the ITS, for example (see Chapter 1, p. 15), in the Australian context, NA groups are perhaps the most closely influenced by deep ecological principles, including the tenets of bioregionalism. These groups have, however, been more focused on establishing localised ethno-communities than on other political philosophical tendencies of land and place connection. Moreover, they reject anarchism's political–philosophical ideal of egalitarianism, advocating instead for hierarchical communities governed by substate nationalist and authoritarian–masculinist principles (Macklin 2005; Sunshine 2008). As the below investigation explores, ideological and practical organisational principles characteristic of international NA movements, such as the advocacy of autonomous 'whites-only' communities, continue to be present in the propaganda and recruitment strategies of contemporary neo-Nazis, and in wider discussions among the far-right social media base, in Australia as well as globally (Taylor 2019; Jackson 2020).

Understanding the uses of 'eco-thought' by far-right groups in Australia also requires acknowledging that proto-fascist movements have always been syncretic (Griffin 1991; Eatwell 1992). The groups and individuals in question use elements of syncretism, 'pastiche' or 'bricolage' (Pisoiu 2015), which have also been a longstanding feature of proto-fascist movements internationally, evident, for example, in the environmental concerns expressed by ENR actors, particularly those with youthful 'identitarian' connections (Zúquete 2018; Valencia-García 2020). In this context, it is important to acknowledge that, since the final stages of the Cold War, right-wing movements with explicit ecological–environmental elements have superficially adopted popular counter-cultural ideas originating from the left, such as global anti-capitalism, anti-imperialism and punk autonomism (Ross and Bevensee 2020; Moore and Roberts 2022).

Mirroring the behaviour of far-right social movements globally (Richards 2022), the Australian examples analysed in this discussion also capitalise in syncretic fashion on the convergence of social, economic and environmental crises. During the COVID-19 pandemic, the ideological environment in which these groups were operating was characterised by the viral spread of misanthropic ecofascist messaging, with members posting phrases such as 'bees not refugees' and 'humans are the virus' shared on cross-national social media (Allison 2020; Beran 2021; Skauge-Monsen 2022). In the US, Australia, the UK and elsewhere, accelerationist and resignatory attitudes towards climate change have also been linked in the context of COVID-19 to the longstanding far-right trope of the decline of the Occident, or 'West', and a specific Fall event symbolising the perceived loss of a Heideggerian sense of Being (Lukács 1952; Beiner 2018). The quasi-Nietzschean response from far- and extreme-right actors to this situation now, as in the past, has involved the overt valorising of national Virtues, the Will to Power, and providence or destiny, which are also broadly connected with far-right accelerationist and survivalist aspects of 'climate crisis' ideological narratives (see Chapter 5, p. 171). Tropes of 'blood' or 'race' then feature as part of a desire to restore cultural destiny or greatness 'beyond Good and Evil' through means of palingenesis, signifying in Griffin's (1991) terms the bloody rebirth of states along racial–national lines (see also Beiner 2018).

Also relevant in this context is that eugenicism and genocidal population reduction measures advocated in far-right responses to climate change are not limited to exegeses on political violence. They also appear as elements of broader international far-right tropes around 'remigration', often among the ENR, US alt-right and identitarians, referring to policies for the forced repatriation of regular and irregular migrants from European settler-colonial cities and towns to their alleged ancestral homelands (Zúquete 2018). That the logic underpinning these proposals is morally dubious is evident from the fact that the development of Global North and wealthy states was historically reliant on the extraction of human labour and resources from less developed countries (LDCs), while those countries now bear the disproportionate detrimental impacts of global heating (Richards 2020). 'Fortress' approaches to borders and migration have, moreover, long applied to the island state of Australia, including in the 70-year history of the White Australia policy since Federation in 1901, and the criminalisation and dehumanisation of asylum seekers in the country's offshore refugee processing programme (McAdam 2013; Crock et al. 2012; Gerard and Weber 2019).

While environmentalist ideas propagated by the contemporary far and extreme right in European colonial societies respond to current crisis conditions, they are also inspired by the enduring influence of various ideological figures and texts. Among several other examples are Savitri Devi's commitment to 'love animals', Ted Kaczynski's urging to 'hate technology' and Pentti Linkola's condemnation of humanity as 'enormously destructive' and claim that human rights are 'a death sentence for all Creation' (Macklin 2022, 983, 987). The neo-Malthusian endorsement by various right-wing actors of reducing human populations in response to global heating has also often been justified by dystopian portrayals of mass migration from the 'Third World' to the West, as outlined in texts such as Jean Raspail's 1975 *The Camp of the Saints* (Ross and Bevensee 2020; Hernandez Aguilar 2023).

In the context of analysing the 'mainstreaming' function of ecofascist and related far-right environmentalist propaganda, it is essential to briefly acknowledge that some texts that resonate with far- and extreme-right groups were not created within ethnonationalist or authoritarian frameworks but contain elements vulnerable to far-right co-optation. As discussed in Chapter 1, the 1968 book *The Population Bomb*, written by Stanford biologist Paul Ehrlich, has been influential in shaping both progressive environmentalism and right-wing climate responses. The book starts with Ehrlich's 1966 visit to Delhi, where he lamented 'overpopulation', despite Paris having a significantly larger population at the time (Mann 2018; Yakushko and De Francisco 2022). Garrett Hardin's theory of 'lifeboat ethics' followed a few years later, arguing that wealthy countries have the moral right to restrict international aid, prevent the poor from accessing their resources and oppose immigration (Hardin 1974). The securitisation of Global South populations by various Global North governments and other multilateral actors has, moreover, also been encouraged by realpolitik narratives about scarcity driving unending cycles of state violence (Sklair 2000), which in turn have been influenced by the overpopulation narratives set out in the Club of Rome's 1972 publication *The Limits to Growth*, despite that this text has also been drawn upon for progressive political ends.

Lastly, the forthcoming discussion implicitly reflects upon the impact neoliberal counter-intellectual environments have on the circulation of far-right narratives and discourses about climate change. In both Australia and internationally, counterfactual narratives about the causes and effects of global heating are encouraged by the monetisation of viral media, which exploits politically extreme and psychologically compelling content for commercial purposes (Gerlitz and Helmond 2013; Powell et al. 2018). This media environment fosters a lack of nuance in debates on critical social issues by promoting clickbait, sensationalism and fleeting patterns of engagement (Charkawi et al. 2021). It is also characterised by a dialectical relationship with legacy media, which similarly at times has a tenuous connection to the 'truth' or factual information about climate science, due to vested economic interests fuelling climate change denialism and the concentrated corporate ownership of media sectors (see Chapter 4, p. 125). Over the past several decades, neoliberal information environments have furthermore contributed to the cultural atomisation of vulnerable internet users (Giroux 2020) while cultivating instrumental and commercial values applied to the notion of 'truth', including scientific truth about climate change (c'Ancona 2017; McIntyre 2018).

Lastly, it is important to recognise the historical origins of contemporary antiintellectualism in far-right media spaces, as well as the anti-knowledge politics surrounding climate science and pseudoscientific notions of 'race'. Proto-fascist 'might is right' attitudes emerged across Europe during the interwar years as an extension of Romantic and reactionary responses to both the cultural progressivism in socialist institutions and the positivistic sociological influences of the Enlightenment period (e.g. backlash against the scientific-spiritual fanaticism of Auguste de Comte, inspired by the positivist, scientific philosophising of Henri de Saint Simon; Eatwell 1996; Richards and Jones 2023). In today's counterfactual and anti-intellectual media environments, sometimes referred to as part of the 'post-truth' era, the neoliberal commodification of knowledge merges with Nietzschean perspectivist influences on the far right (Beiner 2018), resulting in 'truth' being valued by far-right actors only for its political-material use (Biesecker 2018; Richards et al. 2022).

White supremacist attacks

Before discussing the engagements with climate change and environmental politics by far-right Australia-based actors on cross-national media, it is important to account for the global ideological context from which this media has emerged, including the increasing incidence of white supremacist mass violence. On 15 March 2019, Australian-born Brenton Tarrant entered two mosques in Christchurch, New Zealand, and shot to death 51 worshippers (Every-Palmer et al. 2021). A wellknown example of pseudo-environmentalist discourse used by white supremacists was the so-called manifesto by Tarrant published to the imageboard website 8chan before his attack (Cosentino 2020; Moreau 2021). The attack was livestreamed to Facebook in the semblance of a first-person shooter videogame but was removed by moderators 12 minutes after the broadcast ended, by which time it had been viewed by at least 200 users (Besley and Peters 2020; Chai and Yu 2021, 24). The title of Tarrant's manifesto, *The Great Replacement*, clearly alludes to Renaud Camus's populationist theory (Macklin 2019; Wilson 2020). Identifying as an 'ecofascist', Tarrant also incorporated pagan, esoteric Nazi symbols like the *Sonnenrad* on both his person and in the manifesto. The text incorporated references to nineteenth-century German *Wandervögel* and Eurasianist political tropes that associate a nation's culture and political identity with specific geographies (Moses 2019).

Demonstrating his local political connections in Australia, Tarrant had praised the well-known Australian neo-Nazi activist Blair Cottrell as his 'Emperor' and declined an invitation to join the Lads Society from Thomas Sewell, who was at the time the head of the Australian NSN (McGowan 2020). Tarrant was also linked to the pan-European identitarian youth group Generation Identity, having made several donations to one of its core leaders, Martin Sellner (Richards and Jones 2023), as well as to Les Identitaires, the youth wing of French political party the National Rally (Macklin 2019). His manifesto included statements expressing ecological-environmental concern, such as:

> This stripping of wealth and prosperity in order to feed and develop our cultural competitors is an act of civilization [*sic*] terrorism resulting in the reduction in development and living conditions of our own people for the benefit of those that hate us.
>
> (Tarrant 2019)

The massacre at Christchurch was praised in its aftermath by other racist murderers. These figures included a mass shooter in El Paso, Patrick Crusius, whose manifesto stated: 'If we can get rid of enough people, then our way of life can be more sustainable' (Crusius 2019). Philip Manshaus, who fired on a mosque in Norway in 2019, also praised Crusius and described himself as 'chosen' by 'Saint [Brenton] Tarrant' (Burke 2019).

These connections notably occurred within a broader recent history of white supremacist mass violence driven by genocidal or exterminationist racism. While our account is not comprehensive, other pertinent examples include Robert Bowers, who faced federal charges related to a 2018 shooting at a Pittsburgh mosque, where 11 people were killed and 7 injured. Other incidents involve the murder of UK Labour Party Member of Parliament Jo Cox in 2016 by a man with ties to the English Defense League, Thomas Mair, who also yelled 'Britain First' during his attack, and Darren Osborne's vehicular attack at Finsbury Park, near the Muslim Welfare House, 100 yards from Finsbury Park Mosque in London in 2017, which resulted in one fatality and nine injuries. Additionally, a shooting by John Earnest at Chabad of Poway synagogue near San Diego in April 2019 led to one death and three injuries (Richards and Wood 2020).

One of the latest in this series of white supremacist attacks was Payton Gendron's alleged mass shooting in a Buffalo supermarket in 14 May 2022. The shooting was livestreamed on the streaming site, Twitch, until the broadcast was interrupted after almost two minutes and the poster suspended from the platform (Harwell and Oremus 2022). Gendron allegedly posted a 180-page manifesto outlining his motivations. Recalling Tarrant's manifesto, this text also cited Camus's Great Replacement theory. Featuring significant excerpts from Tarrant's manifesto, the document included the passage:

> Green nationalism is the only true nationalism. There is no conservatism without nature, there is no nationalism without environmentalism, the natural environment of our lands shaped us just as we shaped it. We were born from our lands and our own culture was molded by these same lands. The protection and preservation of these lands is of the same importance as the protection and preservation of our own ideals and beliefs. For too long we have allowed the left to co-opt the environmentalist movement to serve their own needs. The left has controlled all discussion regarding environmental preservation whilst simultaneously presiding over the continued destruction of the natural environment itself through mass immigration and uncontrolled urbanization, whilst offering no true solution to either issue. There is no Green future with never ending population growth, the ideal green world cannot exist in a world of 100 billion, 50 billion, or even 10 billion people. Continued immigration into Europe is environmental warfare and ultimately destructive to nature itself. The Europe of the future is not one of concrete and steel, smog and wires but a place of forests, lakes, mountains and meadows. Not a place where English is the de facto language but a place where every European language, belief and tradition is valued. Each nation and each ethnicity was molded by their own environment and if they are to be protected so must their own environments.
>
> (Anon 2022)

The supermarket shooter's manifesto conveyed an exclusivity narrative that echoed the 'replacement' theories (often interpreted as white genocide) promoted by politicians and media figures in Europe, the UK and the US, such as Nigel Farage, Tucker Carlson, Viktor Orban and Donald Trump (Abbas et al. 2022). Its far-right environmentalist dimensions echoed pseudo-Eurasianist themes of land and place connection, anti-industrialism and support for civilisational populationist measures. This included criticism of declining birth rates among white communities and the selective misappropriation of population genetics research. For example, in addition to the above excerpt, the manifesto referenced the work of 38-year-old Michael Woodley, who co-authored a book alleging that ethnicity and cognitive abilities are linked, and authored papers asserting that humans can be divided into subspecies. A table in which Woodley compares humans with animal species such as jaguars and leopards also features in Tarrant's manifesto (Pronczuk and Ryckewaert 2022).[4]

Stormfront

The initial part of our analysis centred on the employment of climate change politics by Australia-based users of Stormfront, the first dedicated online white nationalist

platform (De Gibert et al. 2018). Recognised as one of the most enduring and influential white supremacist websites (Hartzell 2020), Stormfront offers an array of
subforums covering various topics. Among these are informal 'journal clubs' that
categorise population genetics research based on their alignment or opposition to
white nationalist beliefs (Price 2018a) and forums hosting nation-specific discussions, with local branches and subforums targeting site users in Canada, Europe,
South Africa, Latin America, Australia and New Zealand, among other countries
(De Koster and Houtman 2008; Daniels 2009).

Stormfront's pseudo-democratic features, such as its 'Opposing Views Forum',
facilitate a process of negotiated collective identity building (Caren et al. 2012),
where geographically dispersed participants come together as social movement
online communities, creating an imagined white identity community through its
performance and construction online (Caren et al. 2012). Stormfront Downunder,
a nation-specific subforum for Australia and New Zealand, focuses on domestic
political issues and pro-white activism in the region. This subforum is divided into
several Australia-based subforums: General Downunder Discussion, Downunder
Newslinks and Politics & Activism in Downunder.

These Downunder forums retain a white nationalist focus, allowing users to
engage in intergroup debates and build consensus on relevant social and political issues of concern to white supremacists in Australia. In one demonstration of
this, Bliuc and co-authors (2019) analysed 14 years of communication between
Stormfront Downunder users about the so-called Cronulla Riots in Sydney on 11
December 2005. That event involved approximately 5,000 people staging violent
protests at the Sydney beach, displaying signs with phrases like 'ethnic cleansing' and 'take Australia back' (Johns 2017), leading to physical attacks on people
of Middle Eastern appearance, as well as some reciprocal violence from the targeted groups (Richards 2019). Bliuc et al.'s investigation revealed how the event
generated 'unprecedented' levels of activity on Stormfront, contributing to a new
direction for the white supremacist activism of Australian Stormfront users. This
new direction emphasised anti-Muslim prejudice, which gained prominence both
nationally and internationally after 2001 and the US-led war on terror, alongside
resurgent public displays of antisemitism (Bliuc et al. 2019).

For our analysis, we performed quantitative keyword searches and qualitative
thematic coding across 998 posts from the three Stormfront Downunder forums
between 19 June 2008 and 28 February 2018. Users' engagement with climate
change was identified by searching for terms such as 'climate', 'heating', 'warming', 'environment' and 'nature'. All posts were manually reviewed to ensure that
no relevant data were excluded. To expand the dataset, the same word searches
were applied using the advanced search function on the entire Stormfront website,
adding the terms 'Australia' and 'Downunder'.

The majority of the posts on climate change in Australia within the analysed
threads were centred around the 2019–2020 Black Summer bushfires, during
which approximately 24 million hectares were destroyed, 33 people died directly,
and 450 additional individuals passed away until mid-2022 due to the effects
of smoke inhalation (Binskin et al. 2020; Humphries 2022; see also Chapter 4,

p. 127). Numerous threads and comments across Australian Stormfront forums concentrated on prominent counterfactual right-wing news media narratives conveyed both during and in the immediate aftermath of the disastrous event.

One version of this narrative claimed that the fires were not a result of human-induced global heating but were mainly caused by arsonists. Promoted in mainstream news media and by bots on Twitter using the hashtag #ArsonEmergency, in most iterations this narrative suggested that cases of arson were committed by environmental activists to advance a climate change agenda (see Mocatta and Hawley 2020; Weber et al. 2020; see also Chapter 4, p. 128). News articles referencing the theory shared across the threads included, for instance: 'For liberals, it's more fun to blame climate change than arsonists for Australian fires – but arrests show it's not so simple' (Wilson 2020); and 'Those "Climate Change" fires in Australia? Many of them Were Set by Arsonists' (Murphy 2020; SARIAH 7/1/2020). The latter article quoted decontextualised statements made by Dr Paul Read, the co-director of the National Centre for Research in Bushfire and Arson, and contained misleading statements such as 'nearly 85 percent of the fires now burning in Australia have been started, either accidentally or on purpose, by humans' (Murphy 2020).

Other flawed explanations for the fires provided on Stormfront attributed them to the supposed influence of the Australian Greens political party on controlled burning as a technique to manage fuel loads in forested areas, despite the Greens having minimal impact on controlled burning policy and generally supporting the measure. Unfounded accounts of people obstructing emergency mitigation measures were reinforced by posters incorporating statements by far-right Australian politicians, with SARIAH approvingly quoting the ONP's Pauline Hanson as (incorrectly) blaming 'legislation' for preventing people building houses from 'clear[ing] the trees' (SARIAH, see 'Pauline Hanson on Today denies the climate change impacted bushfire crisis', 1/12/2020).

At other times, the fires were explained across the platform with reference to right-wing 'counter-jihad' narratives about groups such as Islamic State and Al-Qaeda that allegedly exploit climate change to harm Western targets. In Stormfront posts, links were provided to articles in support of the counter-jihad conspiracy theory that were primarily published on alt-right US conspiracy websites, including: 'ISIS Tells Followers to Set Forest Fires in U.S., Europe' from The Washington Free Beacon (Kredo 2019), 'Muslim Teen Accused of Starting Aussie Grass Fire Laughs As He Leaves Court on Tuesday' from Gateway Pundit (Hoft 2020), and, erroneously given the country in focus, 'Military-aged Iraqi migrant arrested in Austria following series of arson attacks' from the Geller Report (Geller 2019). Many articles were shared by the prominent Stormfront user SARIAH, who lists their location as 'East Coast Australia' and has 10,576 posts at the time of writing. Citing supposed examples in the US, Israeli, Australian and European contexts, SARIAH described the conspiracy as 'fire jihad', with the fires having been started in an 'Islamist attack on the country' (13/9/2020). While some research has identified Al-Qaeda's propagandised advocacy for 'ember bombs' or 'forest jihad' by starting wildfires in Europe, America and Australia (Fighel 2009; Asaka 2021), there is in reality little evidence of the attempted or successful execution of such attacks by jihadists (Marsden et al. 2014).

Climate change denialism related to Australian bushfires was also a prevalent topic of discussion on the platform more broadly. Coldstar, a 'sustaining member' (that is, a financial contributor) to Stormfront, who has 55,056 posts, reflected the tone and focus of much of the discourse on the fires, amplifying SARIAH's post with their comment: 'It could be the climate change fanatics trying to mislead us or perhaps Muslims who have in the past threatened non-Muslim nations with such actions' (7/1/2020).

In a similar vein, most of the posts on Downunder forums in our dataset pertaining to climate change expressed support for Australian politicians voicing general climate denialist views. Former Senator Fraser Anning, who on 14 August 2018 in the Australian Parliament called for a 'final solution' to Australian immigration (Karp 2018), was praised in several posts for downplaying global heating, as well as for his purported support for agricultural and fossil fuel industries, which are important to some rural Australian communities (McFadden 2022) – though as Chapter 4 elaborates, the respective interests of miners and farmers are in reality often in conflict (see p. 134). These lines of argument tend to echo a historical trademark expression for the Australian nativist right, constituted by an appeal to stereotyped anti-metropolitan and anti-cosmopolitan sentiments among regional and rural communities (Moore 1995). Similar tropes are also apparent in the far-right 'left behind' narratives about regional communities in the US and the UK (see Forchtner 2019c; Rhodes et al. 2019; see Chapter 6, p. 203).[5] For example, a sustaining member of Stormfront, Reynoldsg, with 36,192 posts, initiated a thread titled: 'Australia's Victor Orban Comes Out Fighting'. This thread argued that the 2019 Australian federal election would be contested over issues of 'taxation, climate change, and inequality', expressing the view: 'Fraser Anning, born 1949, has been a senator for Queensland since 2017. Very attached and strongly rooted in his home state of Queensland and Australia, he stands as a defender of Queenslanders and its rural communities' (12/4/2019).

Other prominent commenters, such as LesPatterson, a sustaining member with 5,224 posts, whose location is listed as 'Terra Australis', connected climate change action to centre-left politicians like the ALP's Daniel Andrews, the Victorian Premier at the time of writing, since 2014. In reply to a post titled 'The sound of crickets', LesPatterson wrote: 'So your [sic] a Dan Andrews fan? Going to vote labour [sic] at the next election. I suppose you also believe in climate change as well'. In the same post, he also irrationally linked the Victorian state government with 'the CCCP, the Chinese Communist loving Clive Palmer and his Socialist United Australia Party (UAP), and the Jewish influenced western media' (20/11/2021) – Palmer being a reactionary right-wing mining magnate, former member of the Australian House of Representatives and leader of the far-right UAP (Rimmer 2022). Another post by LesPatterson with input from several others criticised the 2022 incumbent federal ALP government under Prime Minister Anthony Albanese. In an extended passage, authors in the thread railed against institutional-governmental actions to reduce carbon emissions, associating this in coarse terms with high taxation, high welfare expenditure and the devastation of the agricultural industry. These notional policy positions were also expressed by means of racist, anti-Indigenous, homophobic and anti-left invective, in a passage that stated:

> WTF is wrong with people voting for the Greens who have gained seats in the Lower House and Senate, they are just a bunch of homo's, pedo's & trannies who want to open the flood gates for every nonwhite … The green movement is nothing but a communist front, green on the outside and red on the inside. They care for the environment so much that they want to increase immigration out of site, which has huge impacts on the environment and our quality of life. They said that Australia was stolen and want an aboriginal parliament. Go to their website and it looks like something that AOC would write. It's all about closing down fossil fuels, taxing what's left of our industries, closing down farms and giving out trillions in handouts. If it was all implemented we would need to take a wheel burrow full of money to buy a loaf of bread and eat it in the dark.
>
> (LesPatterson, 'Australia's new prime minister Anthony
> Albanese', 22/5/2022)

Some statements on the Stormfront forums examined referred to pride in perceived connections to land and territory. Many of these examples once again associated the 'degenerate' influence of metropolitan areas and liberal city elites with the degradation of the natural environment (see Skauge-Monsen 2022; Chapter 2, p. 47). A Stormfront user named 'AustralianWN', who had 868 posts and whose account is now disabled, for example, lamented the gentrification of Australian coastal areas. In the post 'How money and global exposure are changing the face of Byron Bay' (AustralianWN, 16/7/2021), they state: 'in terms of natural features, it's beautiful, but now like inner Melbourne or Sydney in terms of people, it's just the worst of what this country is, from both extremes of the tosser spectrum, whether it's uber rich or grubby hippy/druggie types'. SARIAH echoed a similar relatively mainstream right-wing colonial trope in Australia, emphasising the connection between their pride in place and their perceived right to inherited privilege, based on their European ancestors' colonial conquering of Australian lands (see e.g. Furniss 2005). Drawing spurious moral equivalence between the experience of British convicts settled in Australia since the late eighteenth century and African slaves transported to America, SARIAH stated:

> My ancestors came on the First Fleet and shortly afterwards. I am proud of their achievements to turn a wilderness into a productive sustaining environment. I will be watching the fireworks at Pyrmont in celebration. In years past I enjoyed visiting Kurnell for the landing enactment and of course exploring the Rocks area. I am grateful to have been privileged to be born in Australia. They were some of the toughest people to have lived. To endure the conditions that they did, and then to go on and build a country in such harsh conditions is astounding. In particular, the convicts among

them were stolen from their homelands, transported to the other side of the world, tortured, put into slave labour camps, subjected to terrible disease, malnourished, etc. etc. but they kept going and achieved great things. Never though have I heard calls for reparations for the ancestors of convicts as is the case for the African-Americans.

('Australia Day Celebrations', 1/4/2018)

Hatred towards Indigenous people and denial of their right to manage the so-called Australian lands and territory were also prominent in the discussion across the threads. A user named 'ctrl', for example, in December 2017 responded to a thread titled 'What are aboriginals and Maori like to deal with?' with extreme prejudice. The views expressed in this case echo latent racist statements more common within Australian society than political leaders truthfully acknowledge (Augoustinos et al. 1999). The violence and moral repugnance of this perspective is evident in ctrl's unabashed endorsement of a continuing colonial genocide against Indigenous people, including through their statement:

A*** are just the worst man, you can't trust them, can't turn your back on them. I learned they have a lower IQ than white people which didn't surprise me at all. They're not the noble primitives living off the land that many would have you believe. They are good for nothing, stupid, violent and lazy. Give them anything and in 3 months they'll have completely wrecked it. The sooner they all die off, the better.

(ctrl, 17/12/2017)

Echoing another popular trope among far-right activists who draw on pseudo-bioregionalist ideas, some posts supporting a notion of 'White Australia' advocated for the creation of localised substate ethnocentric communities. These examples are in some respects reminiscent of the autonomous modes of social organisation advocated in the 2000s by right-wing NAs (Macklin 2005; Sunshine 2008), or the 'back to nature' modes of living advocated by would-be autonomous collectives of far-right actors living in regional outposts, such as the Tasmanian highlands (see Chapter 2, p. 51).

The advocacy of autonomous communities on Stormfront also directly corresponds to the efforts of Paul Innes, the former moderator of the generalist Stormfront Downunder forum, to establish a whites-only community in the Perth foothills, which was discussed in more detail in Chapter 2 (see p. 51). Ozzie Bob, for instance, proposed what he called a 'Stormfront of the street' or 'whites-only living space', referring to local communities where white people live in close proximity to businesses offering cultural facilities and services, consciously supporting their cultural and political survival with a public, low or even 'invisible' profile (Ozzie Bob 2/8/2014). According to his description, and echoing Innes, these communities were inspired by both H Michael Barrett's 2001 notion of 'Pioneer Little Europe' and Friedrich Ratzel's 1901 concept of *Lebensraum*, or 'living space'. It is important in this context to recognise how *Lebensraum* in both world wars was based on the Social Darwinist belief that German people (or in reality the German state) had the right to seek colonial capitalist expansion across Eastern Europe, justifying their violent military actions and social programmes (see e.g. Darré's blood and soil journal, *Odal*; Curthoys and Docker 2001; Staudenmaier 2011a). To justify 'whites-only' communities in Australia, Ozzie Bob wrote:

> Why PLE's for Australia? With the genocidalistic removal of the Immigration Restriction Act 1901 (http://www.foundingdocs.gov.au/item.asp?dID=16) – White Australians have slowly, and now ever increasing, watched Australia descend toward racial chaos and subsequent oblivion, unless Whites take it upon themselves to design their own future. This descent has been assisted direct through Australian Government sanctified and Pre-dominantly Non-White Mass Media Propaganda machines, encouraging Miscegenation and corruption of morality, even including the escalation of Non-White criminality for profit.

To the extent that concerns about global heating in the Australian context appeared on Stormfront, such discourse was primarily found in the generalist forums rather than on the Downunder platforms. For example, a prominent site user by the name of Revision extensively blogs about the threats of wildlife extinction in Australia and the need for conservationist measures to protect the natural environment, specifically emphasising the threat posed by natural disasters and extreme weather to koalas. As a sustaining member of the site with 107,505 posts at the time of writing, Revision's concern is also often articulated within the context of antisemitic, anti-'globalist' conspiracy theories that serve a white supremacist agenda (see also Richards et al. 2022). For example, a post titled 'Koalas Should Be Given Endangered Listing, Environment Groups Say' (26/3/2019) links to an article from *The Guardian* with the same title, which Revision cites appreciatively. However, below this article link are additional links, including one to a 'Holocaust Deprogramming Course'.

Facebook

Discussions about climate change are prevalent on mainstream Australia-based social media platforms such as Facebook and Twitter. But although many of these conversations may contain right-wing nativist or ethnonationalist components, white supremacist online content in both general and far-right Australia-oriented spaces does not primarily concentrate on climate change. The propagandistic use of environmentalist politics in explicitly far-right spaces can nevertheless offer insight into the context surrounding Australian white supremacist narratives. With this in mind, our investigation of Facebook focused on an explicitly white nationalist Australian page, that of the Australian Natives Association (792 followers), with background consideration of other pages and accounts it is linked to, such as Australian Ecofascist Memes (approximately one thousand followers), Offensively Australian (approximately four thousand followers) and the Australian Workers

GANG (approximately one and a half thousand followers). Each of these pages extensively reference well-documented statements from past Australian statesmen, and the ideology of the historical Australian Natives' Association (ANA) from which the Facebook group derives its name. For the purposes of this investigation, all posts on the Australian Natives Association page since its inception in on 1 October 2018 until 29 April 2023 were subject to examination, as were complementary posts across the affiliated pages.

The sites and groups in focus collectively tend to advocate a distinct form of 'radical right' perspective, outwardly dismissing extra-institutional revolutionary violence while claiming to represent the original, mainstream political ideology and institutional framework established for Australia (Hage 2012; Mudde 2019). Unlike Stormfront, which displays neo-Nazi and other overtly genocidal political messages, the far-right propagandists examined here convey a simplistic 'return to tradition' narrative that in reality supports an extreme right-wing socio-political model based on racist exclusion and ethnic cleansing (Mann 2005).

In light of the mainstream and high-profile nature of Facebook, it is necessary to recognise how the platform differs from other sites examined in this chapter, partly by virtue of controversies related to Facebook monetising content associated with political violence. The platform faced criticism for displaying advertising material on Britain First's Facebook pages after the 2019 Christchurch mass shooting by Brenton Tarrant, which was livestreamed via the site (Dearden 2019). It has also been criticised for its 'surveillance capitalist' function of monitoring and curating user behaviour (Zuboff 2019). For instance, the large-scale psychological operations conducted by Aggregate IQ and Cambridge Analytica through Facebook influenced voting patterns in the 2016 UK Brexit referendum and the 2016 US presidential election (Risso 2018). In these cases, far-right supporters of Trump and Leave political lobbyists used the platform to disseminate unsubstantiated claims of a connection between microeconomic labour conditions and international migration (Scott 2019). In the context of the present discussion, it is also relevant that Facebook has been known to host numerous influential conspiracy theories; around 4.5 million QAnon conspiracy proponents were active on Facebook and Instagram in August 2020, at the height of the COVID-19 virus pandemic, spreading white supremacist, antisemitic and conspiratorial narratives about the disease (Wong 2020).

Given its commercial, sensationalist and reactionary right ideological environment, Australian far-right actors have successfully leveraged Facebook, using the site to connect with a wide audience base, and cross-fertilising in propaganda networks with international far- and extreme-right groups. The former Australian neo-fascist organisation the United Patriots Front (UPF) had 120,000 followers on its Facebook page before being removed in May 2017, allegedly at the request of Australian security agencies. In a 2018 interview, former UPF leader Blair Cottrell stated, 'Facebook's been extremely effective for us; it's, ah, indispensable to the development of our organization' (Richards 2019, 47); while on the site, UPF page administrators shared anti-immigration and xenophobic propaganda that mirrored many of the narratives and tactics of the European far-right collective Generation Identity and the former neo-Nazi Greek political party Golden Dawn (Richards 2019). Far-right anti-Islam rallies organised by UPF predecessor Reclaim Australia were also coordinated through Facebook, attracting thousands of attendees in 2015. Recently active extreme-right groups on Facebook have also included UPF's descendant organisation, the Lads Society, as well as, briefly, other neo-fascist and neo-Nazi groups like Nationalist Alternative, Nationalist Uprising and Soldiers of Odin (Peucker and Smith 2019).

The ANA Facebook page is distinct from more openly far- and extreme-right Australia-based pages, which have mostly, by the time of writing, been heavily censored or removed. Claiming to want only to return Australia to its white nationalist origins, the group takes its name from and claims continuity with the large and influential early Australian fraternal organisation of the same name (McKenna 1996). By the statements on its website, this group aims to re-establish the ANA organisation as a race-based self-help body, with small grants for business and individual hardship, as well as art and essay competitions with cash prizes (ANA 2022). Combined with the white nationalist slogan 'Our Own for Our Own', its central interest reflects the early historical focus of the ANA on the private social welfare sphere that defined its raison d'être (ANA 2022).

It is important to note, then, that the original ANA organisation ceased to exist when its last remaining operation – an insurance company – merged with Manchester Unity in 1993 to form Australia's largest insurer, Australian Unity (Bindon 2016). The ANA was indeed one of the most influential among several mutual fund or 'friendly societies' in Australia, having been established in the nineteenth century, with membership restricted to 'white' people of European descent born in Australia. Members would pay a regular subscription fee in exchange for the ANA covering their wages if they fell ill, supporting members' families if they were unable to work, and covering funeral expenses for members and their families (Menadue 1971). The organisation, along with the RSL, was notable for supporting the establishment of the White Australia policy in 1901, and its membership included Australia's first Prime Minister, Edmund Barton, and later Prime Ministers including Alfred Deakin, James Scullin and Francis Forde (Denheld 2011; National Museum Australia 2023). The ANA is now officially defunct, with limited contemporary political influence, and its last remaining branch in Western Australia closed in 2007.

Also relevant is that the ANA page on Facebook that we examine here was established by Matthew Grant, a well-known far-right activist in Australia who spoke at Reclaim Australia and UPF rallies, coordinated with the New Zealand-based neo-Nazi group Action Zealandia, and was a former leader of the Australian extreme-right Eureka Youth League, the youth division of Jim Saleam's Australia First Party. Just prior to the establishment of the ANA, Grant set up the White Millennial Society, a Canberra-based fraternal organisation that attempted to forge relationships, belonging and common identity between young white people (see e.g. El-Khoury 2015). Despite the ANA's outwardly nonviolent stance, Grant was recognised by local antifascists as endorsing the accelerationist neo-fascist Iron March site, which was linked to several racist murders, and whose members have described themselves in conversation with NSN recruiters as having 'nationalist socialist' political commitments (WhiteRoseSociety 2023). Controversy also surrounded Grant and the ANA in 2023, when the organisation successfully registered with the NSW Government as an approved hunting organisation, such that at the time of this book's writing, its members may legally obtain firearms licences (McKenzie and Galloway 2023).

Although the Facebook followers of the current ANA are relatively few in number, Matthew Grant's activities and the existence of the group in the wider ecosystem of the Australian far right demonstrate the influence that the original ANA's legacy, and representations of Australian history and tradition, continue to have on the contemporary Australian far right. Correspondingly, the ANA Facebook page primarily features straightforward memes combining quotes and portraits of respected mainstream Australian statesmen (Australian Natives Association 2022). It is important to note how these figures, who include John Curtin, Alfred Deakin, William Spence, and founders of the white nationalist Australian Workers' Union, are celebrated within Australian culture as 'forefathers', while they are also widely recognisable to everyday Australians due to their commemoration in street names, universities, local councils and political electorates (see Figure 3.1; see also Chapter 2, p. 35, 50). ANA quotes featured on Facebook referring to such figures also invariably reference the White Australia policy and its defence. In this reading, 'White Australia' is represented as a non-negotiable element of Australian society and a source of cultural and social stability, and even survival (Fitzgerald 2007).

The ANA page and the Australian Ecofascist Memes page to which it is linked also promote a distinct Australian brand of white supremacy, claiming allegiance

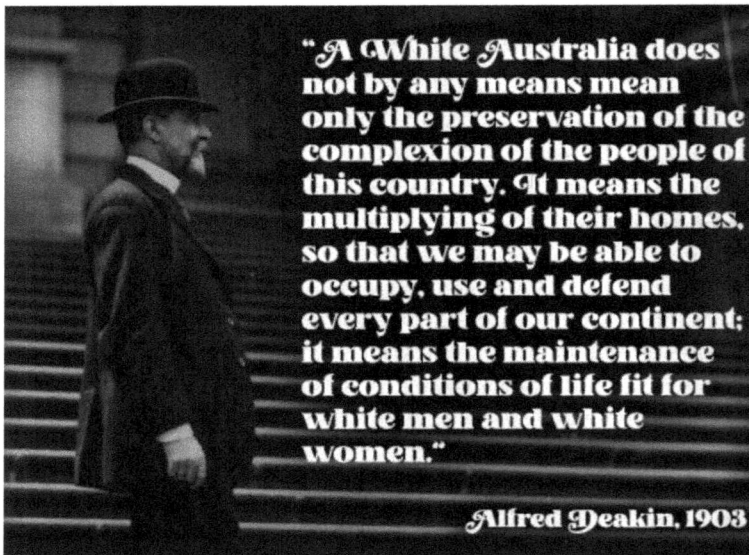

Figure 3.1 ANA Facebook page 15 December 2021

to a populist, worker-based identity and the strong early historical endorsement of a 'left-wing', labourist politics of Australian white supremacy (Griffiths 2006). Australia is thus seen as the 'white workingman's paradise', with the White Australia policy primarily designed to protect 'battlers' – or non-elites – against the competition from foreign workers brought into the country to serve the greed of exploitative capitalist elites (see Chapter 2, p. 39).

Across the ANA feed, a unique Australian blend of 'patriotic and socialistic' values, which are portrayed as inseparable from a cultural ethnic heritage (e.g. 'Australian Nativists Association', Facebook, 18/1/2022, 12/4/2019, posts in 2023) might be said to reflect a 'national socialist' or 'red-brown' political orientation (Ross 2017). In some ways, this side of ANA propaganda recalls the tendencies of National Bolshevism, or, more directly, the ethno-states and 'ethnocultural separatism' advocated by Aleksandr Dugin's *The Fourth Political Theory* (2012). This is not to suggest convergence between disparate further left- and right-wing positions, as in Faye's (2002) 'horseshoe theory', but rather, we suggest that it signifies a Strasserite tradition, pointing to the syncretic and utilitarian deployment of 'socialism' by contemporary far-right actors, including those active in Australia, for the purposes of entryism and recruitment (Ross 2017).

The ANA's anti-neoliberal position also seeks to garner working-class support by combining nationalist and racial ideologies under an umbrella of social conservatism. This propaganda tactic appears to oppose the most extreme facets of capitalist systems, without genuinely challenging their foundational principles (see Ross 2017). Foreshadowing the later rise of national socialism and fascism in Europe, the historical advocates of White Australia, who are celebrated by the contemporary ANA, also incorporated a strong antisemitic element, presented as an opposition to 'international capitalists'. While this sentiment was common in early twentieth-century Europe, it manifested in many mainstream political literary works in Australia at the time, including *The Money Power* (1921) and *The Kingdom of Shylock* (1917) written by Frank Anstey, an Australian Labor MP from 1910 to 1934. Demonstrating the enduring influence of this history, a link

AUSNATIVES.ORG

i

Australian Cossacks – The Australian Natives Association

Australian Cossacks 22nd April 2020 General Globalisation in parallel to the influence of modernity has persistently since colonial time...

Figure 3.2 ANA Facebook page 22 April 2020. Accompanying text: 'We wish to reach the point where the average person could think of an Australian and in their mind; picture the kinds of lives being lead [sic] by ANA members, dedicated to hard work, the development of their family clan, connection to soil and place, endless resolve and firm character: carrying the torch and the heritage left to us by our forebears'.

to an electronic copy of Anstey's *The Money Power* is periodically shared on the ANA Facebook page.

The ANA Facebook page commemorates anniversaries based on the original organisation's 1871 founding, reinforcing the current page administrators' assertion that the contemporary ANA is over 150 years old ('Australian Nativists Association', Facebook, 17/1/2021). The original ANA was, in fact, one of the initial promoters of a national day honouring colonial conquest on 26 January, pre-dating the official Australia Day, which the current page administrators dub 'ANA Day' ('Australian Nativists Association', Facebook, 26/1/2022; see Chapter 2, p. 35). This historical framing allows the ANA and its wider associated social networks to present themselves not as a radical alternative to existing systems, but as adherents to the natural ruling ideology most compatible with Australia's current political institutions. For supporters, their perspective is the 'true' Australian view, only displaced in recent decades, and therefore, it might also be seen as entirely reasonable and non-radical. This is supported by meetings held by the Facebook page administrators under the ANA banner. For instance, at the 2022 'National Convention', a photo showing the backs of 14 members – perhaps signifying David Lane's '14 words'[6] – mostly wearing plaid, was accompanied by the following statement:

> On the evening of the 151st year of the Australian Natives' Association, The Canberra & Melbourne chapters of the organisation met once again, as is now tradition on the banks of the Murray river to reflect on the victories of the association and consider the path ahead for the coming years.
>
> ('Australian Nativists Association', Facebook, 30/4/2022)

Similar to most ultranationalist viewpoints seen in Australia, the ANA page asserts a distinct identity rooted in the unique Australian landscape. Quotes from historical Australian statesmen and other notable colonial figures reference the 'wattle clad hills of Australia' ('Australian Nativists Association', Facebook, 26/1/2021; see also the article on the ANA webpage linked in this post, and a 1905 Henry Lawson poem titled 'Waratah and Wattle' shared on Facebook on 1 September 2020), emphasising the supposedly great struggle to shape the particularly inhospitable Australian environment towards productive agriculture. This notion of the fit and able-bodied man is reminiscent of both the historical Australian narrative of the rascal nation and its people developing into virile maturity and independence as one (Curthoys 1999; Veracini 2007). In this respect, the Australian myth might also be said in some respects to echo the belief in fascist Italy that a nation's populace can experience a spiritual resurgence, spurred by the peasantry's conquest of nature (Armiero 2014). In both propagandised narratives of national identity, overcoming the severe environment and cultivating the land symbolises the people's collective strength, determination and potential for revitalisation.

A contradiction within far-right narratives that glorify European colonisation stems from the environmental damage caused by colonial agricultural activities in Australia and elsewhere. Claims of settler-colonial entitlement to land ownership

Figure 3.3 ANA Facebook page 30 December 2021

in Australia often go hand in hand with assertions that Indigenous peoples merely 'wandered the land' before colonisation, without actively shaping or engaging with it in a way that maximised productivity (McNamara 2014; Sparrow 2022). This narrative implies that Indigenous people neither occupied nor defended the land, and that it was only through significant settler efforts that the land became hospitable – a blatant falsehood that disregards Indigenous histories of sustainable land management, and their steadfast resistance against conquest and dispossession (McKenna 2002). The ANA also rationalises its claims of land ownership by emphasising Australia's military history, including the Anzac 'diggers'' defence of their homeland discussed in Chapter 2, in service to the British Empire during WWI, and in resistance against Japanese aggression during WWII (ANA Australia Day 2021 blog post linked on FB).

As also explored in Chapter 2, narratives of settler-colonial struggles, featuring tales of miners and convicts, along with an Australian legend emphasising perseverance and hard work, have historically played a significant role in Australian far-right and white supremacist propaganda. Mythic notions of Australian identity that ideologically supported the right to clear land were often rooted in Burkean narratives of *terra nullius*, which posited that a territory was unoccupied in order to justify colonial occupation (Lindqvist 2007). Such narratives are now widely recognised as contradicting histories that acknowledge Indigenous peoples' extensive and careful management of the Australian environment to sustainably cultivate a diverse and abundant ecosystem that supported a large population (McNamara 2014; O'Brien 2016). These historical accounts reveal that the land colonial settlers first encountered was not a 'wilderness', but rather a complex, managed environment that was systematically destroyed by settlers in order to establish a more labour-intensive and environmentally damaging system of European grazing-based agriculture (Rose and Australian Heritage Commission 1996).

In some ways similar to the incursions the ALR made into organic and non-organic farming communities after WWII, the obvious ecological features of the identity of ANA include its celebration of land improvement and wilderness management projects, constituting what the organisation sees as foundational elements in its land ownership claims (Gaynor 2012). These activities are also associated with a notion of national culture and identity predicated on rural life and work, set in contrast to 'urbanism' (see Figure 3.4). The ANA is also, however, resistant to any consideration of broader environmental issues, including climate change. Except for the superficial promotion of national parks and regional rivers as sites for conferences and places for wholesome leisure activity, there is no evidence of meaningful engagement with actual environmental issues on ANA channels or other platforms they are linked to.

While the ANA outwardly expresses a nonviolent political stance, leaders of the organisation hold connections to violent groupuscules such as the National Socialist Network (NSN), as previously mentioned, and the Facebook page is also connected to more overtly far-right or extreme-right pages, such as Australian Ecofascist Memes, now an inactive page, with its last post dated 18/7/2022. The Australian Ecofascist Memes page primarily hosts either incomprehensible or crude racist memes, or memes and posts from the ANA page, which also mostly

The truest wisdom in the world is to encourage men not to settle in the comfortable luxuriance in the cities of civilisation, but to go into the wilds, where your fathers went before you, and laid the foundation in hardship, suffering and toil of the prosperity you are enjoying today.

The Honourable George Reid 1903

Figure 3.4 ANA Facebook page 29 May 2020

contain racist remarks made by colonial Australian figures. The posts on the ANA page and other pages to which it is networked exhibit a racist opposition to East Asian migration, often described as a 'Yellow Peril' narrative, which was also prominent in earlier periods of Australian coloniality. The ANA page hosts, for example, an image of Andrew 'Banjo' Paterson warning on 31 August 1901 of 'danger' from an 'Oriental invasion' (see Figure 3.4). Another post from the ANA that was shared across several networked channels features an adapted Eureka flag with the captions 'Fuck Urbanites' and 'Meme Brought to You by the Australian Workers Gang', echoing a white nationalist sense of rural republican-labourism, also prominent in Australia's political history (see Chapter 2).

While the concept of the Great Replacement (Camus 2012) holds sway in modern far- and extreme-right circles, as discussed in Chapter 2, its core themes have long been a cornerstone of the racist or 'racialist' Australian perspective embodied by the ANA. At the heart of this media lies the perception of a threat of being 'swamped' by Asian immigration, as expressed, for example, by Pauline Hanson in her first speech in the Australian Parliament in 1996 (Australian Human Rights Commission 2021; see Chapter 2, p. 32). Fear of the potential decline of the white population and traditional family structures is a prevalent source of anxiety for the nativist Australian right. This anxiety also extends to fear of white demographic

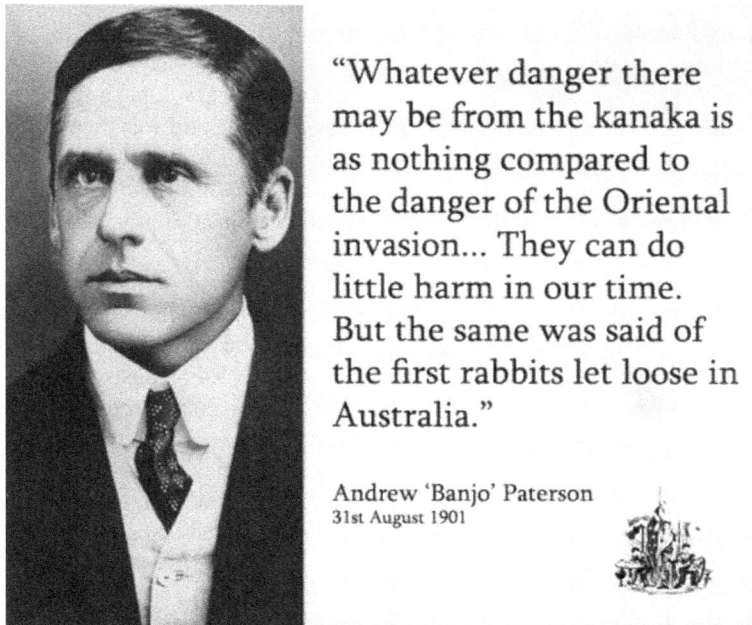

"Whatever danger there may be from the kanaka is as nothing compared to the danger of the Oriental invasion... They can do little harm in our time. But the same was said of the first rabbits let loose in Australia."

Andrew 'Banjo' Paterson
31st August 1901

Figure 3.5 ANA Facebook page 25 November 2018. Accompanying text: '"Banjo" Paterson's poetry and stories, still popular to this day, are among the best writings of our national culture, often evoking a strong affection for, and affinity with, the Australian bush and community'.

90 *Ecofascism online*

collapse and threats non-white workers ostensibly pose to the living standards of whites, and the national and international capitalist class – often coded in far-right media as Jewish – are often presented as driving these threats (Busbridge et al. 2020; Hanebrink 2018).

Gab

For the next part of the analysis, we examined Australia-oriented discourse on climate change across the white supremacist platform Gab. Established in 2016 and launched in May 2017, Gab is an alternative technology platform known for its lax moderation and limited restrictions on contentious, hate-filled or provocative speech (Jasser et al. 2021). The more stringent moderation of mainstream social media platforms like Facebook and Twitter, prior to Trump's significant Twitter activity between 2017 and 2021, and Elon Musk's radical deregulation of the platform after he assumed ownership in October 2022, inspired the development of new platforms like Gab. Gab and other 'alt-tech' platforms such as Parler and Trump's platform Truth Social were established with a view to catering to reactionary right-wing audiences. Gab's features thus resemble a combination of Facebook and Twitter features (Jasser et al. 2021), wherein users can create dedicated pages and share public messages via a 'gab', similar to a tweet, which initially had a 300-character limit (Jasser et al. 2021) but then increased to 3,000, and now is 5,000 characters.

To analyse the discourse about climate change in Australia shared among a more generalist far-right audience, we examined the Gab Australia page (handle: g/australia), which had 73,900 followers at the time of writing and was created on 6 May 2018. The number of members on the platform surged following Tarrant's mass shooting in Christchurch in 2019, increasing from 4,500 in March to 11,000 in June 2019, and reaching over 45,000 members in March 2021 (Guerin et al. 2021).

Some research has been conducted on Gab Australia, including its key influences and content. A study hosted at the Centre for Resilient and Inclusive Societies (CRIS) found that 'far-right mobilisation' on Gab Australia was driven primarily by a few highly active accounts. In their analysis of 45,404 posts between January and September 2020, CRIS researchers found that only three accounts were responsible for 75% of the content (Guerin et al. 2021), while key topics of interest included COVID-19, 2020 pandemic management measures in Victoria, Australia, Black Lives Matter, China–Australia relations, and climate change. Many discussions across this dataset centred on conspiracy theories involving Marxist, socialist or communist plots to infiltrate institutions, such as the media, schools and universities, allegedly, according to platform users, for the purpose of destroying Western civilisation and establishing a communist–socialist (and anti-white) governmental regime (Guerin et al. 2021). Further research based on this dataset revealed the importance of mainstream media, with left-leaning outlets being criticised, and conservative or right-wing sources being used to legitimise reactionary or prejudiced beliefs. News Corp's television channel Sky News

Australia and the online tabloid *Daily Mail*, owned by the British-based parent company Daily Mail and General Trust, were key sources referenced by posters, with nearly one-third of all links across the threads directing to these outlets (Peucker and Fisher 2022, 16).

Another study analysing 393,430 Gab Australia posts across three one-month periods during 2019 revealed its focus on international politics and events, with narratives reflecting former US President Trump dominating US-centric content across Gab, also indicating the place of this politics in the wider far-right ideological landscape. The study's link analysis showed that Gab users often accessed what researchers called 'countermedia' (Droogan et al. 2022), or far-right alt-news sources, as they are referred to in Chapter 4 (see p. 139). For the purposes of this investigation, it is important to note that Gab has notably historically hosted pages managed by and associated with violent neo-Nazi Australian groups, such as Antipodean Resistance and the Nationalist Australian Alternative (Campion 2019).

In this research we sought to explore the discussions of climate change on Gab among a generalist, Australia-focused far-right audience. For the analysis, we collected all posts on the page between 16 August 2022 and 7 February 2023, isolating those that mentioned 'climate change', 'global warming' or 'global heating', resulting in 734 posts. Among them, 'climate' appeared 1,127 times, 'warming' 227 times and 'heating' 13 times, with 'climate change' mentioned 867 times. A word cloud generated in NVivo, displaying the top 100 words that featured recurrently across all 734 posts, is shown below:

After NVivo was used to conduct word frequency coding queries on the 734 posts we scraped from the Gab Australia page, a thorough qualitative analysis of a random sample of 100 posts was performed. Through a close thematic reading of the posts, the primary topics that emerged were revealed to be similar in theme and nature to some of the content across Stormfront, including posts related to denialist conspiracy theories questioning the credibility of climate change science, and others undermining meaningful measures to reduce GHG emissions. Out of the 100 posts examined in depth, only two suggested acceptance of the reality of climate change, its human-driven causes or its negative impacts. Of these, one post advocated for keeping wolf spiders as pets, claiming that they can help reduce global heating, and encouraged users not to kill them; the other expressed scepticism that climate change reparations should be paid by developed to less developed states. Most of the conspiracy theories shared on the platform had nationalistic and reactionary white supremacist undertones, or overtones, focusing on the alleged nefarious motives of global climate change action advocates, including among governments, scientific organisations and community climate activists.

Many posts linked climate change to other contemporary conspiracies, such as QAnon-related theories which argued that the COVID-19 virus was not real, and that government surveillance and control measures instituted as a means of managing the pandemic were in reality applied to gain greater, tyrannical control over national populations. Much of this discourse mirrored the wider prevalent ultranationalist and racially charged interpretations of the pandemic as a 'hoax' perpetrated by either China or an international elite. One post titled 'The Great

Figure 3.6 Word map generated in NVivo showing the 100 most prevalent words on the 734 posts from the Gab Australia page, with erroneous 'stop words' excluded.

Awakening' (Robert55, 24/11/2022), for example, includes the statement: 'More vaccines, more surveillance, more "Gain of Function", and all in the name of Climate Change'. The title of this thread referred to the 'awakening' component of the QAnon conspiracy theory, which contends that a secretive international network of Satanic child exploitation is controlled by Jewish global(ist) political and commercial elites and that the Gates Foundation is using COVID-19 vaccinations to microchip people for mass surveillance (Busbridge et al. 2020). In another anti-semitic display, on 18 October 2022, TJAntipodes used the moniker of a capital-ised 'They' to signpost a Jewish-controlled cabal of globalist elites.[7] Alleging a conspiracy that changing weather patterns are actually attributable to Earth passing through a 'galactic plasma current', this user posted:

> They will reveal this to us slowly in the next few years, what They have known and started Davos in 1971 to deal with. All their amoral hatred with the covid inoculations and actions are driven by getting us under control and diminished by 2030 so we don't disturb their amoral dark future.

Other contributors outlined generic conspiracies that have long been articulated to explain climate change-related weather patterns. These theories often suggest that the weather has been purposely manipulated by corrupt governments to deliberately control or enslave global populations, with the term 'geoengineering' appearing 37 times in the dataset. One poster, ORIGIN8 (22/10/2022), claimed that the Victorian Premier Daniel Andrews used 'weather modification patents' to justify 'cloud seeding', which modified the weather in order to cause widespread flooding on Australia's East Coast in 2022. In another similar post, titled 'Trump Nukem', a user going by the name of Ragnarokrag linked to the NSW and Victorian government websites, highlighting what they termed the 'Rain-Making Control Act 1967' and the 'Snowy Mountains Cloud Seeding Act 2004', citing these documents as evidence for their claim that: 'Australia Legislation and Companies Using Cloud Seeding/Chemtrails for Geo-Engineering Climate Change' (20/11/2022). Other arguments alleging governments' intentional orchestration of global heating effects were occasionally made without any explanation. For instance, IvyBelle (6/9/2022) wrote: 'NO, YOU'RE NOT CRAZY & NEITHER IS MOTHER NATURE … ALL THIS CRAZY "CLIMATE CHANGE" WEATHER IS INTENTIONAL & BY DESIGN'.

Allegedly corrupt elite media institutions were also at times said to be behind public political campaigns pushing for greater climate change action, with several Gab users citing the supposed authority of alt-news websites in their place. One user called Susanishere, for example, linked to the Australian conspiracy website Tott News, posting on 20 October 2022, 'Melbourne appoints Rockefeller-backed "chief heat officers" … Melbourne City Council has appointed "chief heat officers" to deal with the "life-threatening" effects of climate change'. The national public broadcaster, the Australian Broadcasting Corporation (ABC), was often blamed; for example, Sutherlanach stated: 'Thus, the cooling is good news for crazy warmists. I expect the ABC to tell us all about it and reduce the levels of climate anxiety, especially among teens … I'm sure they shall. As soon as Hell freezes over' (7/11/2022). Opposition to climate change politics were also often connected to the far-right reaction to progressivist politics. In this line of reasoning, Penguinite wrote: 'See, indoctrination does work! Our ABC and it parent BBC will be empowered at this result. Just look at the way we are being force fed Aboriginal culture. They know adults are more discerning and recognise 1984 brain washing techniques but kids not so!' (26/11/2022).

These and other posts in the dataset might be said to exemplify a 'post-truth', anti-intellectual perspective, characterised by emotive and sensational misinformation combined with the partial coverage of social issues (McIntyre 2018). The conspiracy narratives shared by Gab users were frequently not of their own creation, however. Instead, they often drew on denialist conspiracies promoted in alt-news media, or anti-progressive agitating spouted in conservative Australian media or international alt-right publications such as Breitbart and RT (see Allcott and Gentzkow 2017; Lazer et al. 2018; Vosoughi et al. 2018). In this context, 'post-truth' might be understood as an expression of ideological supremacy on the part of political and economic elites, based on their dismissal and disregard of established knowledge and evidence, usually with the aim of reinforcing social hierarchies or advancing political agendas (Biesecker 2018).

The anti-intellectual politics evident on Gab targeted climate scientists and academics. For example, one post on the page criticised 'a team of psychological researchers from Australia [that] is delving into the mindset of those who refuse to buy the notion that man-made climate change is quickly creating' (FuriousFolly 24/9/2022). Others described reports produced by the Australian Bureau of Meteorology (BOM) and the Commonwealth Scientific and Industrial Research Organisation (CSIRO) as 'Propaganda Dressed as Science', produced by 'organisations drifting into … political advocacy' (Penguinite, 7/12/2022). Some posters targeted academics in a manner we suggest is reminiscent of the anti-intellectual politics of proto-fascist organising in previous eras; exemplary here is Barnabasmo's comment: 'Bizarre "Fact Check" of World Climate Declaration Claims No Natural Climate Change for Almost 200 Years … Academics around the world are becoming increasingly frustrated and angry at the politicisation of science in the interest of promoting the command-and-control Net' (1/9/2022). The Australian Academy of Science (AAS) was also a prominent target, with another post by T_J1776 on 1 September 2022 stating: 'In a submission to the Australian Government, the AAS has demanded "disinformation" about climate change, the great barrier reef, and Covid vaccines be censored … Watts Up With That?'

Continuing the criticism of the credibility of scientific claims underlying climate change politics, Tim Flannery, who won the 2007 Australian of the Year for his work as a scientist, conservationist and climate change activist, and an award for his 2005 book, *The Weather Makers*, was described in several Gab posts as 'mentally ill'. Ozhomeschool criticises Flannery's award after 'a number of his claims were known to be false … The $90 million Government grant was also given after his terrible record for facts and truth was well known' (3/2/2023). Activists were frequently targeted as well, with Sasseybritches linking to a Sky News Australia article and sarcastically stating: 'Genuine climate scientists …Astro physicist have been saying we are going into a solar minimum for years. How is that global warming going Greta? Lucky she is the greatest scientific mind in history' (3/11/2022).

Resistance to effective climate change policy was sometimes legitimised by users through reference to certain political authorities, such as conservative right-wing politicians in Australia. Several users in particular appreciatively highlighted Sky News Australia's reporting on a clash between ONP leader Pauline Hanson and ALP Minister for Foreign Affairs Penny Wong over the ALP's introduction of some limited policy measures to reduce GHG emissions (e.g. FuriousFolly, 8/9/2022). Another poster by the name of ozpat shared YouTube videos of ONP Senator Malcolm Roberts instructing a school student that climate change is not real (9/9/2022). Lozzo1 then linked to a page on Senator Roberts's official site, titled 'So it's come to this – Globalist push on Climate Change', urging other users of the website to 'vote against' a world where climate change is not real, but hunger and poverty caused by human activities are used by elites as a means of control (9/9/2022).

Non-far-right politicians in Australia were sometimes criticised for their positions on foreign policy or the local economy in relation to climate change.

Barnabas22, for example, shared material from the far-right lobby group Advance Australia, citing an alternative media article that stated: 'Beijing is preparing for war while Canberra worries about climate change' (6/12/2022); while BankWarrior argued that under the ALP's energy policy, Australia will become 'totaly [sic] dependent on China for wind our and solar power' (19/12/2022). Several posts also raised the potential consequences of the post-2022 ALP Government's shift away from fossil fuel exports for the national economy and living standards. BankWarrior claimed that 'Albo and Chrissy Bowen' are intent on wasting money on the 'Climate Change' issue (7/1/2023). Referencing an alternative media article from Electroverse, ozhomeschool emphasised pride in Australia's colonial history, connecting this sense of nationalism with a warning about the supposed dangers associated with the government's net zero agenda:

> The Aussie government is advancing its dangerous Net Zero agenda … The coal-rich nation is making its cheap and reliable energy – the backbone of its modern civilization's success – expensive and unreliable. Welcome to fuel poverty, Australia – a fast-track back to pre-industrial times, a throwing away of everything you – and the collective we – have built over the past two centuries.
>
> (8/10/2022)

The alleged adverse economic impacts of climate change mitigation measures on residents of regional towns throughout Australia and New Zealand were frequently discussed in the dataset, including often with reference to certain policies of New Zealand's Labour Party instituted under Jacinda Ardern's leadership (2017–). This government introduced a tax on agricultural producers for the methane emissions produced by their livestock, a measure that aligns with recent findings from the IPCC (2023) and other international scientific bodies about methane emissions from industrial agriculture driving global heating. One poster stated, 'Farmers need to stick it out with their protests against WEFS, stooge, Arden and her Fart and Burp tax. Long live the farmers' (Ditchyoursmartphones911 21/10/2022). Another wrote, 'The government's plan, the He Waka Eke Noa proposal, will devastate farmers, small-town life, and the country's agricultural industry' (Lightntruth 15/10/2022); while one user shared a video titled 'New Zealand to Introduce Cow "Burp" Tax to Tackle Climate Change' and commented, 'Politicians will continue to "do as they please" as long as the people allow it' (soundtree 14/10/2022). Linking the New Zealand Labour Government's actions to fears of a sense of encroaching liberal progressivism in Australia, another user named Winslayer criticised Ardern, 'Next, this chief of the "tiny island nation" will sue Australia over climate issues! She's a spokesperson for the NOW' (WinSlayer, 29/9/2022). The scientific basis of the policy was also questioned by Sutherlanach, who stated:

> The size of the Australian and New Zealand beef herds are but a small fraction of the size of the cattle herds in both Africa and India respectively …

But only in the Antipodes are those cattle killing the climate … And maybe only here are some farmers gullible enough to go along with this bullshit.

(28/11/2022)

Occasionally, climate change conspiracy theories were linked to other far-right, white supremacist or misogynistic ideologies in a blended manner. This often led to comparisons between the increasing public awareness and acceptance of climate change and non-traditional expressions of gender and sexuality:

While livestock farmers in the Netherlands are being climate change regulated out of existence, school children are being indoctrinated to eat bugs … I personally, am Hyper-Carnivorous, so I would have all the Eco-Spastics Permanently Politically Expelled from my nation … Sangmoore's commentary on the Ethnocidal Predators wanting to reduce meat consumption () to effeminise straight white men [their primary evolutionary threat / threat to their power retention], using the lame & hollow emotional blackmail excuse of 'climate change'.

(FuriousFolly, 23/10/2023)

In some instances, conservative stances on climate issues were linked to unqualified claims that Australia and other affluent Global North nations should not be required by the international community to provide compensation to countries severely impacted by global heating. Conveying this sentiment in connection with a conspiracy theory that rising sea levels are not actually a result of melting ice due to climate change, one user with the pseudonym Oohsheet stated:

'THE SINKING PACIFIC ISLANDS SCAM' … Australia sits on bedrock and pacific islands on coral that breaks down over time due to age, weather, buffeting seas and the weight of man made devices like 30 ton excavators and other building equipment … The island leaders being historical fans of graft and corruption and with the help of globalists then package this as seas rising because of man made climate change and then get paid handsomely by the feelings based politicians in Australia.

(24/9/2022)

Telegram

The final segment of our analysis focused on overtly violent extreme-right neo-Nazi groups in Australia and their engagement with political environmental themes through Telegram. Telegram's moderation approach is relatively lenient, prohibiting only violent and illegal pornographic content shared in public channels (Hughes et al. 2022). The platform has correspondingly hosted far-right users, including ecofascists, and spaces where these groups gather, which are sometimes colloquially known as 'Terrorgram' (Hughes et al. 2022). Terrorgram, also referred to as 'Fascist Telegram', is a collection of approximately 300 far-right Telegram channels shared across various other channels to increase the far right's online visibility (Loadenthal 2022).

While much of the far-right content on Telegram is then accelerationist and implicitly ecofascist, providing context for other far-right informational material, our focus is on the online communications of two recently active Australian extreme-right networks, which have several dozen members nationally: the NSN and the EAM, which are both led by Thomas Sewell. The NSN was established in early 2021, resulting from a merger between the Antipodean Resistance and Lads Society (Knaus 2021), while the EAM was formed shortly after.

The NSN has attracted considerable public interest through widespread media coverage since its inception. In early 2021, multiple news sources documented NSN members partaking in white supremacist chants, performing fascist salutes and participating in cross burning while camping at Grampians National Park in Victoria. In March 2021, Sewell made headlines for assaulting a black security guard outside the Nine Network's Melbourne office after attempting to confront the producers of *A Current Affair*, an investigative documentary series that had featured an episode about the NSN that week (AAP 2022). Public interest reignited when Sewell was arrested in May 2021 on charges of assault, armed robbery and property damage, originating from an incident where he and other NSN members attacked a group of men in a Cathedral Ranges car park (McGowan 2021). In 2021 and 2022, Adelaide-based NSN members faced charges, including alleged offences related to possession of information for terrorist acts and explosives-related offences (ABC News 2021).

Numerous in-depth news articles in late 2021 also discussed a months-long investigation into Australian neo-Nazi organising conducted by journalists affiliated with the Nine Network, with the help of antifascist researchers from the White Rose Society and Slackbastard, and in end-stage consultation with Victorian and Federal law enforcement and intelligence agencies (McKenzie and Tozer 2021a). In this case, a private investigator infiltrated the NSN, secretly recording his encounters with network members and their broader activities in pubs, suburban parks, and at their operational headquarters in a rented property in Rowville, Victoria. The investigator also leaked information from encrypted chat rooms used by the NSN and EAM. In the Channel Nine reporters' words, the recordings collectively exposed a 'cult-like breeding ground for extremists who are training in hope of bringing about societal collapse or a white revolution' (McKenzie and Tozer 2021a). In March 2023, further reporting also detailed connections between Australian Defence Force members and neo-Nazi groups, including the NSN, and the efforts of security agencies to deter politically violent individuals from travelling abroad for military training, following one NSN member's trip to Ukraine to fight alongside neo-Nazi groups against Russia (McKenzie and Galloway 2023).

The data analysed in this section, then, comes from the NSN and EAM Telegram channels, comprising 453 NSN posts and 87 EAM posts between 8 January 2020 and 27 July 2022. The NSN aims to make 'Australian National Socialism a serious political force', echoing the neo-fascist trope of national socialist domination. The EAM embraces Third Positionist ideas of bioregionalism and ethnic separatism, striving to create what it calls 'autonomous White Australian communities' (NSN 8/3/2021).

In reality, the EAM and NSN share almost the same membership. Despite differences in their stated goals, their Telegram pages feature nearly identical propaganda. Thus, they are often collectively referred to in this analysis. Both groups endorse Social Darwinist visions of a natural order, which is proto-typical of fascism (Griffin 1991). By using environmentalist imagery in their propaganda, they also might be said to be seeking support for a version of 'natural law', purportedly reflected in the Australian landscape. Although the NSN is more openly nationalist socialist and genocidal than many of their white supremacist counterparts in Australian history, these tropes indicate a propagandised leverage of the natural 'law of return' ideologically drawn upon by other antisemitic white supremacist groups, such as the interwar proponents of Percy Stephensen's AFM, or Eric Butler, the leader of the ALR, who attempted to recruit from the early post-war organic movement (see Bird 2014; Chapter 2, p. 55).

In both NSN-affiliated Telegram channels, political arguments placing white men at the apex of a dubious natural hierarchy are explicitly framed by associating

Figure 3.7 NSN Telegram channel 19 February 2020. Accompanying text: 'Our land is a primordial expression of the Natural order and accordingly deserves our respect'.

Figure 3.8 EAM Telegram channel 17 September 2021. Accompanying text: 'The brothers in Brisbane have been busy forging their Active Club. True adherence to Natural Order requires us to mould our bodies into examples of human perfection. Get outside and train with your brothers!'.

what it means to be an NSN member with physical activity in natural environmental settings. The text accompanying these images also often incorporates victimisation narratives, similar to the early demographical replacement narratives of Australian colonial actors fearing an 'invasion' from the East (Ang and Colic-Peisker 2022; Wilson 2022). Camus's Great Replacement white genocide thesis, which posits that white populations are being systematically replaced by non-white migrants, particularly African, Arabic and Muslim people, with the complicity of a group of elites (Cosentino 2020), also looms large in NSN material. For instance, one message in the Telegram dataset states: 'The only reason White Australia lies in such a grave situation is because there aren't enough White men resisting the invasion. Stop being a part of the problem and get involved today. Inactivity is death!' (NSN 9/1/2020). Another frames non-white migration to Australia as an existential threat to the 'white race', with the words, 'Get active in the struggle for racial survival. Your brothers are waiting for you' (NSN 19/1/2020).

The NSN's response to the alleged demographic replacement of white people resonates with Stormfront discourse around white-only communities or PLEs.

Their goal of creating parallel white communities in rural–suburban enclaves also builds upon an aspiration previously articulated by NSN leader Thomas Sewell on Facebook and YouTube videos when he headed its predecessor organisation, the Lads Society. In one video interview, for instance, Sewell described the Lads Society as:

> The building block of our future community … A community sort of like, what do you define it as? Like, you know, white Australians, or Aussie Australians, I mean there are so many colloquial terms, we don't actually have a sort of homogenous community with its own interests, with its own community goals. No one is really looking after us.
>
> (Mattys Modern Life)

Demonstrating a degree of consistency in this ideology, the administrators of the NSN Telegram channel publicly declared their intention to '[build] a network of White Australians across every city, suburb and town who are against the systematic replacement and destruction of White Australians' (NSN 12/3/2021).

In addition to the political history of White Australia, the contemporary Great Replacement theory has also shaped these narratives. Stemming from other white genocide theories prevalent in the extreme right worldwide (e.g. the 'Zionist Occupation Government' theory) (Kivisto and Rundblad 2000; Greene 2019), the Great Replacement holds that political and economic elites are deliberately replacing white populations in wealthy states with people from LDCs, particularly Muslims, through outbreeding and large-scale migration (Camus 2012).

Although 'replacement' ideas are sometimes, perhaps debatably, considered less extreme than earlier conspiracy theories, the concept is frequently exploited by violent white nationalist and white supremacist activists. Moreover, in this context it is important to note that the theory has been gaining purchase in mainstream political arenas as well as on the extreme right. For example, naturalised Australian citizen Lauren Southern, who now has a regular commentary spot on Sky News Australia, popularised the core theory in activist missions she participated in to disrupt humanitarian NGO boats from rescuing migrants in the Mediterranean Sea, and in a documentary she produced about post-apartheid South Africa (Richards and Wood 2020). Great Replacement rhetoric has moreover long been echoed in European identitarian groups' propaganda and mass media references to far-right activity. Further, in 2021, Fox News host Tucker Carlson suggested the existence of a plot led by the Democratic National Committee (DNC) to promote the demographic 'replacement' of white Americans for the purpose of creating pro-globalisation DNC voting blocs in the country (Miller-Idriss 2022).

Similar to the general white nationalist discourse on Stormfront and radical right nativist community-building on Facebook through groups like the ANA and Australian Ecofascist Memes, claims to land and territory across the NSN channels also frequently refer to Australia's settler-colonial history. Both Telegram channels positively cite the radical transformation of Australian lands through white settlement and the oppressive control of Indigenous peoples. Using imagery such as Captain Cook (see Figure 3.9), the act of settlement is argued to make people of European descent in Australia the rightful inheritors of Australian lands. This narrative is occasionally intertwined with borrowed ideas from Aryan mythology (see Whitsel 2001) portraying white European settlers as 'Aryans' who arrived in Australia and liberated the land from what white supremacists describe as 'savage' Indigenous communities (McKenna 2012).

Claims to land and territory that are exclusivist and supremacist are further emphasised on the NSN Telegram channel through the dual constructs of whiteness and masculinity. The narrative promoted by the group's channels establishes whiteness through lineage, tracing back to the arrival of the First Fleet in Australia, which founded a convict settlement at Sydney Cove on 26 January 1788 (Calma 2015). While this colonial history is celebrated in propaganda images, the portrayal of NSN members participating in physical activities like hiking also evokes

Figure 3.9 NSN Telegram channel 26 January 2020. Accompanying text: 'January 26th, 1788 – the birth of Aryan civilisation on the Australian continent. It is crucial that we commemorate the deeds of our ancestors and furthermore, ensure that their toil was not in vain'.

Figure 3.10 EAM Telegram channel 23 August 2021

the cultural–historical legend of the 'Australian bushman'. Depicted as a down-to-earth, free-spirited and masculine figure, the larrikin Australian archetype, also present in NSN propaganda images, prioritises physicality over intellectual pursuits, resisting both individualism and external authority (Ward 1965). In this context, an Other – typically represented by non-whites – is excluded by the homogeneity of the Australian man. At times, this Other is portrayed by adapting the Australian myth through the lens of a quasi-republican-workerist viewpoint, as demonstrated by the NSN's dismissal of both imperial and capitalist (referred to as 'globalist') historical personalities (see Murrie 1998, 68).

As explored in Chapter 2, the fundamental mythology surrounding Australian settlement systematically excluded individuals of non-European origins, particularly targeting Chinese and Indigenous populations, as well as women, from the Australian 'fraternity'. Within the NSN, related idealised concepts of white masculinity emphasise the embrace of aggression, a Nietzschean Will to Power and Darwinian ideas of survival of the fittest (Lukács 1952). These ideological stances are communicated across both Telegram channels using visuals and language that endorse conflict and aggression. Themes of dominance and submission are displayed through NSN members' activities such as boxing, mixed martial arts, outdoor training, hiking and other forms of homosocial bonding where recruits interact with fellow white men. Male bonding holds significant value within this instance of far-right political organisation, as it enables members to collaboratively shape their masculinity (Köttig et al. 2017) and appeal to potential new members (Kimmel 2007).Figure 3.12.

Hiking in natural settings is particularly significant for NSN recruitment, drawing historical parallels with the *Wandervögel* youth movement of the Weimar

Figure 3.11 NSN Telegram channel 21 August 2020. Accompanying text: 'Not since the days of Rhodesia have short shorts looked so good on racists. It's time to bring them back into fashion. The Sydney boys will be at the centre of that. It's a huge transformation for the movement to make and NSN will be resented because of our leading role, but without that transformation, White Australia will not survive'.

period, which was largely absorbed into the Nazi movement during the 1930s. As Biehl and Staudenmaeir note, the *Wandervögels*' 'back-to-the-land' emphasis spurred a passionate sensitivity to the natural world and the damage it suffered (1995, 10) as a result of industrialisation.

While the NSN does not, by comparison, explicitly express concern about environmental issues related to climate change, several propaganda examples illustrate the strategic use of environmental politics for propagandistic purposes. For example, on 24 March 2021, NSN Queensland activists posted photos of themselves rescuing a 'cold, abandoned Bandicoot joey' during an overnight camping trip. They also advocate hiking, with notable instances including an

Figure 3.12 EAM Telegram channel 30 September 2021. Accompanying text: 'Combat sports are vital to the future of Australian National Socialism. Practice of them gives us a warrior skillset, keeps us fit and nimble, and provides men with confidence to act as men'.

event at the Cathedral Ranges north-east of Melbourne in May 2021. During this event, they were accused of robbing and assaulting other hikers whom they labelled as 'Antifa', suspecting them of filming the group (AAP 2021). In other demonstrations, such as those in Victoria's Grampians in January of that year, NSN members sang the Australian folk song 'Waltzing Matilda', wore black t-shirts featuring Celtic symbols, chanted 'white power' and burned a cross in a ritual reminiscent of the Ku Klux Klan (McKenzie and Tozer 2021b). Other similar hiking activities are also routinely advertised in EAM posts, as in the following statement: 'NSW EAM tested their cardio by hiking a 7km section of Sydney's Coastline. Join your local chapter, tribe and train' (EAM 4/7/2022). A description of training exercises conducted for recruits was also shared on the EAM channel on 26 June:

Discipline and persistence is our greatest strength, constancy over the years has confounded the opposition while we grow through adversity.

The EAM recognises that a group is greater than just the sum of its individuals, and we stand in solidarity with all other groups of White Men and Women that struggle for a future for our Race.

Training Regime.

1km Run warm up.

1x round Shadow boxing

3x rounds Partner shadow sparring

6x rounds 4 for 4 combo drills

3x rounds Light sparring

3x rounds Heavy sparring

88 burpees to finish up. (EAM, 26/6/2022)

Physical activity in NSN groups' messaging is also juxtaposed with references to subordinated masculinities and the 'black-pilled' mindset, the latter of which signifies a nihilistic dismissal of the possibility for successful white supremacy within certain far-right communities (Preston et al. 2021). The channel administrators offer advice to members on ignoring the cognitive dissonance that may arise from their participation in these activities, stating:

Melbourne/Vic EAM Active Club did a 16.5km hike up 300m elevation with overnighter packs last weekend … Keep it simple, Regular, Group, Activity! The more regular, the bigger the group, the better the activity, the closer we get to victory! Ignore the blackpills, ignore the demoralisers, ignore the critics, ignore all those that say it can't be done, ignore the fear mongers, ignore the cowards, ignore the globo-homo faggot tyranny. Get outdoors with your mates, climb a mountain and take a group photo that drives fear into the hearts of the traitors and racial aliens that occupy our soil. Blood and Honour!

(EAM, 15/3/2022)

As part of the NSN's effort to construct a civilised masculine ideal, its Telegram feeds display a strong opposition to feminism and support for radically traditional Evolian reactions to the perceived gender degeneration the organisation associates with modern liberal capitalist society. As Graff and his co-authors have recently demonstrated, 'gender conservatism' can serve as an ideological bridge connecting various components of the global far right (see Di Sabato and Hughes 2022). In this regard, NSN statements about appropriate social roles for women can be understood in relation to the 'tradwife', or traditional wife, phenomenon.

The contemporary far-right tradwife movement mainly consists of online propaganda shared by both men and women within far-right social movements, advocating for the exclusive and white supremacist embrace of traditional gender roles. This can feature women being responsible for child-rearing and home-making, men financially supporting their families, wives 'submitting' to their husbands, as well as expectations of monogamy and heterosexuality, rejection of feminism and usually the promotion of broader white nationalist agendas to which some of these value systems are connected (Sabato and Hughes 2022). European aesthetics in this propaganda may also signify the ecofascist trope of 'purity', alongside a rejection of the alleged deteriorating influence of modern decadence or despoliation. Paradoxically, the reinstatement of these binary gender ideals is often claimed to provide agency to women who might otherwise be excluded from participating in alt- and far-right circles (Campion 2020). For example, on the NSN Telegram page, it is stated:

> Men are being conditioned by this same liberal capitalist culture to view women as expendable prostitutes. They are told that women are to be judged solely by their ability to provide sex or pornography. This is, of course, to be mediated by the provision of money, whether as actual prostitutes or via apps like Tinder which treat the user as a commodity.

(23/12/2020)

Concerning the natural environment, the limited references to women in the NSN and EAM Telegrams and their statements about gender might also be viewed as a consequence of a broader far-right appropriation of the tradlife and possibly the 'cottagecore' movements. Where cottagecore aesthetics online initially emerged from queer progressivist communities, before they were incorporated into far-right narratives,[8] in this context, tradlife reflects a binary and reactionary worldview. Originating in far-right online circles, tradlife stands for 'traditional life'. Similar to tradwife, it refers to the promotion and idealisation of conservative, traditional gender roles, family structures, and cultural values, rejecting feminism and LGBTIQA+ rights, as well as multiculturalism. The term is often used by those who advocate for a return to what they perceive as simpler and more stable times, with an emphasis on traditional social norms. That worldview tends to promote a rural connection to the land and a rejection of symbolic and physical polluting or desecrating influences on white-only environments, which are said to range from individualism and promiscuity, to television, fast food and non-white migrants. Images of obedient tradwives then often construct the viewer through a male gaze (Farrell-Molloy 2022).

Displaying the influence of radical traditionalist ideology on the Australian far right, women are frequently depicted in NSN propaganda in a traditionalist and binary manner, portrayed as needing rescue and protection. For instance, on 31 January 2021, the NSN Telegram feed included the factually incorrect and violently racist statement: 'Ten n******* gang-raped two fifteen-year-old girls in Brisbane. This is why we burn crosses!' This theme, moreover, echoes a prominent refrain of the contemporary identitarian and alt-right outside Australia. In the late 2010s, for example, in response to a mediatised moral panic about sexual violence allegedly perpetrated by non-white irregular migrants, the UK's Generation Identity displayed 'British Girls Matter' banners, while the Identitäre Bewegung in Austria displayed 'Respect for Women' banners (see Richards 2022).

Figure 3.13 NSN Telegram channel 7 March 2020. Accompanying text: 'Go out and experience the beauty of our fatherland. Only with a true understanding of what we are fighting for, will we have the courage to lead this struggle until its end'.

In March 2023, the NSN organisation further capitalised on traditionalist tropes of 'protecting' women in connection with the historically resonant far-right political campaign targeting transgenderism. In this case, more than a dozen NSN members made Hitler salutes on the steps of the Victorian Parliament in Melbourne, expressing support for an anti-trans rights rally led by far-right UK political presenter Posie Parker. In a livestreamed video, NSN leader Thomas Sewell, in the aftermath of the event, stated to well-known former leader of the UPF Blair Cottrell: 'Here is the evil, here is the absolute chaos, here are all the freaks and scum of society. And then here are these people who have genuine concerns about child grooming, about Safe Schools, about what's going on' (@ PPantsdown 2023). Articulating a fascist vision for capitalising on contemporary anti-trans hatred, through the use of a visual environmental metaphor, Sewell then added: 'And they're not really organised, they're not disciplined, there's no real strength and character to them. They're kind of lost in the wilderness and we want to represent like ah, a vanguard, a spearhead and become a separating force like parting the sea between the two' (@PPantsdown 2023).

Concluding comments

The analysis in this chapter has revealed that Australia-based far-right online media networks exhibit some interest in environmental issues, but this interest is typically limited to a narrow and traditional form of environmentalism, which emphasises a blood and soil worldview and a neo-Malthusian focus on population (see Farrell-Molloy 2022). Such views are sometimes accompanied by a more modern concern about conservation and pollution, but this perspective is typically expressed in terms of certain population groups being viewed as 'pollutants' rather than a broader, holistic (or global) concern for environmental crises affecting soil, oceans and the atmosphere (Orton 2000). Despite commentary on such events as the 2019–2020 Black Summer bushfires across segments of the media under consideration, climate change is not generally viewed as a serious concern by contributors to these online communities (Mocatta and Hawley 2020). In fact, climate change action or even the science itself is most often dismissed as part of a global anti-white conspiracy. Only a very small number of individuals contributing to these channels view climate change as a serious issue, and even then, these individuals are usually subject to criticism and ridicule from their peers.

In particular, the far-right online forum of Stormfront exhibits little engagement with environmentalism, and its users tend to express views in line with a 'natural order' perspective. Climate change is not taken seriously by the community on Stormfront Downunder; but a more limited number of users who positively identify with ecofascism, such as Australia-based Revision (see Richards et al. 2022), contribute extensively to the site's generalist forums and do highlight the harmful effects of global heating.

On Facebook, the ANA group draws on eco-adjacent themes, including through the use of imagery and lyricism that denotes a love for the bush, though the users of that page also demonstrate a lack of serious consideration of climate change. Telegram's NSN and EAM groups, then, are focused on the Australian outback specifically in connection with the fascist ideal of outdoor exercise and observation of what they call 'a natural order'. This community exhibits similarities with traditional Nazi ecofascist positions, valuing masculinity, idealised nature, clean living, physical accomplishment, ruralism, and a 'taming and working the land' model of quasi-Lockean land ownership. While the NSN does not focus on climate change, it uses environmentalist signalling for the purposes of recruitment and branding. This messaging also coheres around themes of bloody race war; for example, posts about movement building are framed by the mission statement that 'this is how we win'.

While aspects of the Australian far- and extreme-right milieu may represent a form of ecofascism that focuses on human 'pollutants' (Moore and Roberts 2022), it is important to note that the ecofascism of mass shooter manifestos – which incorporate an ostensibly serious acceptance of climate change to justify violent action – remains a fringe position. It is also worth noting that mass shooters themselves represent only a tiny fraction of far- and extreme-right actors, although their actions are deeply concerning and demand attention. Indeed, the prevalence of environmental themes in the manifestos of many far-right politically violent actors underscores the importance of examining the role of such perspectives within far- and extreme-right movement spaces and communication platforms.

As yet, there has been little consideration of how such right-wing media and internet subcultural ecosystems contribute to the creation of an accelerationist terrorist like Tarrant, who at once regurgitated a slew of predictable and expected contemporary tropes such as the Great Replacement conspiracy, while also proclaiming an ecofascist perspective that contradicts the overwhelming Australian right-wing consensus on climate change denialism (Macklin 2022). Tarrant self-describes as an ecofascist and in his manifesto made statements suggesting that he takes environmental issues relatively seriously and does not deny the threat of anthropogenic climate change. This represents a position that clearly does not reflect the denialism of the Australian far-right media – to be discussed in the following chapter – or the denialism of the majority of contributors to far-right online and social media channels. We might reasonably conclude, then, that in these contemporary far-right spaces, there is a comparative lack of effective ideological authority and hierarchical transmission of orthodoxy (see e.g. Nissenbaum and Shifman 2017).

As mentioned in this chapter's introduction, historical influences are frequently cited to explain the recent environmental focus of extreme-right violent actors (Campion 2023). These influences may include German scientist Ernst Haeckel's 1866 coining of the term 'ecology', Reich Minister of Food and Agriculture Richard Walther Darré's advocacy for '*blut und boden*' (blood and soil), the inaccurate claim that Hitler was a vegetarian, or the presence of biodynamic farming outside Nazi concentration camps (Biehl and Staudenmaier 2011). However, the most critical aspect of ecological *fascist* groups, as evidenced poignantly in the Australian context, is that the underlying logic of much of the violence in question is centred on the re-establishment of 'white power', not a holistic preservation of ecology. Where holistic ecological preservation emphasises the sustainable interdependence between humans and non-human elements, contemporary ecofascists adopt an interpretation of ecology in which 'we are all connected', but this connection is characterised by death, destruction and domination, predicated on spurious 'natural hierarchies' (Eatwell 1996; Staudenmaier 2021).

The following chapter shifts focus to consider the ideological landscape that gives rise to various right-wing climate responses within the Australian setting. Providing a critical examination of the origins of climate change denialist ideologies, it examines how Global North exclusivity and dominance in relation to the impacts of global heating are maintained and supported by political messaging in Australian news and political media industries. The analysis specifically investigates how denialism, delay and obfuscation in relation to the anthropogenic causes and impacts of global heating are both supported by and persist through these institutions' historical and continued backing of fossil fuel industries.

Notes

1 Griffin (2003) describes how the 'groupuscular right' represents the extreme right's sophisticated transnational adaptation towards metapolitical strategies, away from party political ones, after 1945, which allowed ultranationalist groups to persist despite attempts to suppress them. They are 'fully formed' and 'autonomous' but have capacities to share linkages with likeminded networks who also pursue the palingenetic, revolutionary aims of reforming states along racial–national lines.
2 The *Sturmabteilung*, or 'SA', often referred to as the Brownshirts or Stormtroopers, was a paramilitary organisation affiliated with the Nazi Party in Germany during the 1920s and 1930s. Founded in 1921, the SA played a crucial role in Adolf Hitler's rise to power by providing security at Nazi rallies, intimidating political opponents, and engaging in street battles with rival political factions.
3 This is an environmental philosophy and movement that emerged in the 1970s, largely through the work of Norwegian philosopher Arne Næss. It advocates for a holistic approach to environmental issues and 'biospherical egalitarianism', emphasising the intrinsic and equal value of all living beings and the interconnectedness of nature. Deep ecology goes beyond superficial solutions to environmental problems and calls for a fundamental reevaluation of human societies' relationship with the natural world.
4 Woodley has been affiliated with the Vrije Universiteit Brussel and received degrees from Columbia University and Royal Holloway University of London. Alex Mas Sandoval, a researcher in population genetics, started an online petition after the Buffalo shooting to have Woodley's PhD revoked.
5 The 'left behind' narrative typically refers to the perception that the socioeconomic and cultural decline experienced by certain groups, particularly white working-class individuals, is a result of policies favouring minority populations, immigrants and refugees. This perspective can fuel resentment, hostility or fear towards these groups, and is usually fomented and exploited by politicians or extremist groups to promote a racially divisive and exclusionary agenda.
6 David Lane's '14 words' is a white supremacist slogan derived from a longer statement that he wrote. The 14 words are: 'We must secure the existence of our people and a future for white children'. Lane was an American white supremacist and a leader of The Order, a group responsible for multiple murders. The phrase has become a popular slogan among white supremacists and neo-Nazis, symbolising their belief in the need to preserve the white race.
7 QAnon theory also holds that 'deep state' operatives hostile to Donald Trump are aligned with non-state political actors, such as antifascists in the US, who are also said to be funded and directed by the billionaire Jewish philanthropist (and WWII refugee) George Soros (Amarasingam and Argentino 2020).
8 By comparison, cottagecore – also sometimes called farmcore – is an aesthetic trend that promotes certain activities aligned with traditional life, including sewing, gardening, farming, romanticised regional living and vintage clothing, with these activities engaged in for pleasure rather than as a form of work. Cottagecore first emerged from progressivist queer communities and their online ecosystems as an aesthetic fantasy of escape from the late modern capitalist city grind (Barbeau et al. 2022). As with other left-wing aesthetics and counter-cultural tendencies, such as skinny jeans, hipster haircuts and counter-cultural gaming communities, cottagecore has been appropriated in syncretic fashion, along with the development of 'tradlife', in support of far-right political ideologies (Moore and Roberts 2022).

References

@PPantsdown (20 March 2023) 'Neo-nazi Thomas Sewell, who led the nazi contingent, expresses his ideological alignment with the KJK/Deves/Deeming anti-trans rally. He

just thinks they need some discipline & organisation', *[Twitter]*. https://twitter.com/
PPantsdown/status/1637790288537157632

Abbas T, Somoano IB, Cook J, Frens I, Klein GR and McNeil-Willson R (2022) 'The
Buffalo attack: An analysis of the manifesto'. https://icct.nl/publication/the-buffalo
-attack-an-analysis-of-the-manifesto/

ABC News (8 April 2021) 'Improvised explosives, extremist material found as two men
arrested in Adelaide', *ABC News*, accessed 1 June 2022. https://www.abc.net.au/news
/2021-04-08/two-arrested-over-explosives-extremist-material-in-adelaide/100054938

Allcott H and Gentzkow M (2017) 'Social media and fake news in the 2016 election',
Journal of Economic Perspectives, 31(2): 211–236.

Allison M (2020) '"So long, and thanks for all the fish!": urban dolphins as ecofascist fake
news during COVID-19', *Journal of Environmental Media*, 1(1) Supplement 1: 4–1.

Amarasingam A and Argentino MA (2020) 'The QAnon conspiracy theory: a security threat
in the making', *CTC Sentinel*, 13(7): 37–44.

Ang S and Colic-Peisker V (2022) 'Sinophobia in the Asian century: race, nation and
Othering in Australia and Singapore', *Ethnic and Racial Studies*, 45(4): 718–737.

Anker P and Witoszek N (1998) 'The dream of the biocentric community and the structure
of utopias', *Worldviews: Global Religions, Culture, and Ecology*, 2(3): 239–256.

Anon (2022) *You wait for a signal while your people wait for you* [Due to the sensitivity of
this content we do not include a link to the original document].

Anstey F (1917) *The kingdom of shylock*, Hesperian Press, Perth.

Anstey F (1921) *The money power*, Fraser & Jenkinson, Melbourne.

Armiero M (2014) 'Introduction: fascism and nature', *Modern Italy*, 19(3): 241–245.

Asaka JO (2021) 'Climate change-terrorism nexus? A preliminary review/analysis of the
literature', *Perspectives on Terrorism*, 15(1): 81–92.

Atkins E and Menga F (2022) 'Populist ecologies', *Area*, 54(2): 224–232.

Augoustinos M, Tuffin K and Rapley M (1999) 'Genocide or a failure to gel? Racism,
history and nationalism in Australian talk', *Discourse & Society*, 10(3): 351–378.

Australian Associated Press (AAP) (2021) 'Hikers "terrified" by neo-Nazis who allegedly smashed
car windows as they fled, Melbourne court hears', *The Guardian*, accessed 26 November
2022. https://www.theguardian.com/australia-news/2021/oct/27/hikers-hysterical-when-neo
-nazis-allegedly-smashed-car-windows-as-they-fled-victorian-mountain-court-hears

Australian Associated Press (AAP) (20 December 2022) 'Neo-Nazi Thomas Sewell facing
jail time over attack on Nine Network security guard', *The Guardian*, accessed 1 June
2022. https://www.theguardian.com/australia-news/2022/dec/20/neo-nazi-thomas
-sewell-facing-jail-time-over-attack-on-nine-network-security-guard

Australian Human Rights Commission (2021) *The Racial Hatred Act: Case study 3*. https://
humanrights.gov.au/our-work/racial-hatred-act-case-study-3-0

Australian Natives Association (ANA) (2022) *Australian Natives Association* [Facebook],
accessed 22 November 2022. https://www.facebook.com/AusNativesAssociation

Australian Natives Association (ANA) (2022) *Home*, accessed 22 January 2023. https://
ausnatives.org.

Barbeau E, Blanchard E, Qısın L and Almeida VS (2022) 'Queer fragmentation and trans
urban aesthetics: from cyberpunk to cottagecore', *Glocalism: Journal of Culture,
Politics, and Innovation*, 2022(1): 1–38.

Beiner R (2018) *Dangerous minds: Nietzsche, Heidegger, and the return of the far right*,
University of Pennsylvania Press, Pennsylvania.

Beran O (2021) '"Environmentalism without ideology" and the dreams of wiping out
humanity', *Filozofija i Društvo*, 32(3): 439–459.

Besley T and Peters MA (2020) 'Terrorism, trauma, tolerance: bearing witness to white supremacist attack on Muslims in Christchurch, New Zealand', in Peters MA and Besley T (eds) *The far-right, education and violence*, Routledge. https://doi.org/10.4324 /9781003096788

Biehl J and Staudenmaier P (1995) *Ecofascism: lessons from the German experience*, AK Press, Chico.

Biehl J and Staudenmaier P (2011) *Ecofascism revisited: lessons from the German experience*, New Compass Press, Porsgrunn.

Biesecker BA (2018) 'Guest editor's introduction: toward an archaeogenealogy of post-truth', *Philosophy & Rhetoric*, 51(4): 329–341.

Bindon WP (25–28 August 2016) 'Two enduring treasures: why they are not ornaments', [conference paper], Australian & New Zealand Masonic Research Council Biennial Meeting and Conference, Launceston, accessed 3 November 2022. http://linfordresearch .info/fordownload/Periodicals/ANZMRC%20Proceedings/ANZMRC%20Proceedings %202016.pdf#page=76

Binskin M, Bennett A and Macintosh A (2020) *Royal commission into natural disaster arrangements: report*, Commonwealth of Australia, https://naturaldisaster .royalcommission.gov.au/system/files/2020-11/Royal%20Commission%20into %20National%20Natural%20Disaster%20Arrangements%20-%20Report%20%20 %5Baccessible%5D.pdf.

Bird D (2014) *Nazi dreamtime: Australian enthusiasts for Hitler's Germany*, Anthem Press, London.

Blaustein J, Fitz-Gibbon K, Pino NW and White R (eds) (2020) *The Emerald handbook of crime, justice and sustainable development*, Emerald Group Publishing, Bingley.

Bliuc AM, Betts J, Vergani M, Iqbal M and Dunn K (2019) 'Collective identity changes in far-right online communities: the role of offline intergroup conflict', *New Media & Society*, 21(8): 1770–1786.

Bookchin M (1987) 'Social ecology versus deep ecology: a challenge for the ecology movement', *Green Perspectives: Newsletter of the Green Program Project*, accessed 26 November 2022. http://dwardmac.pitzer.edu/Anarchist_Archives/bookchin/ socecovdeepeco.html

Breivik A (2011) *2083: A European declaration of independence* [Due to the sensitivity of this content we do not include a link to the original document].

Brown CS (2005) 'Ecofascism and the animal heritage of moral experience', *Dialogue and Universalism*, 15(7/8): 35–48.

Burke J (12 August 2019) 'Norway mosque attack suspect "inspired by Christchurch and El Paso shootings"', *The Guardian*, accessed 1 June 2022. https://www.theguardian .com/world/2019/aug/11/norway-mosque-attack-suspect-may-have-been-inspired-by -christchurch-and-el-paso-shootings

Busbridge R, Moffitt B and Thorburn J (2020) 'Cultural Marxism: far-right conspiracy theory in Australia's culture wars', *Social Identities*, 26(6): 722–738.

Byrne G (2020) 'Climate change denial as far-right politics: how abandonment of scientific method paved the way for Trump', *Journal of Human Rights and the Environment* 11(1): 30–60.

Calma T (2015) 'Australia survival day', *AQ-Australian Quarterly*, 86(1): 10–12.

Campion K (2019) 'A "lunatic fringe"? The persistence of right wing extremism in Australia', *Perspectives on Terrorism*, 13(2): 2–20.

Campion K (2020) 'Women in the extreme and radical right: forms of participation and their implications', *Social Sciences*, 9(9): 149–169.

Campion K (2023) 'Defining ecofascism: historical foundations and contemporary interpretations in the extreme right', *Terrorism and Political Violence*, 35(4):1–19.

Camus R (2012) *The Great Replacement*, RWTS.

Caren N, Jowers, K and Gaby S (2012) 'A social movement online community: Stormfront and the white nationalist movement', in Earl JS and Rohlinger DA (eds) *Media, movements, and political change*, Emerald Group Publishing Limited, Bradford.

Carter A (2012) 'Packaging hate: the New Right publishing networks', *Searchlight*, accessed 26 November 2022. https://www.searchlightmagazine.com/wp-content/uploads/2017/05/Far-right-publishers-FebMar2012.pdf

Chai WXT and Yu J (2021) 'Influence of social media on deviant acts: a closer examination of live-streamed crimes', in Khader M, Neo LS and Chai WXT (eds) *Introduction to cyber forensic psychology: understanding the mind of the cyber deviant perpetrators*, World Scientific. https://doi.org/10.1142/12164.

Charkawi W, Dunn K and Bliuc AM (2021) 'The influences of social identity and perceptions of injustice on support to violent extremism', *Behavioral Sciences of Terrorism and Political Aggression*, 13(3): 177–196.

Cosentino G (2020) *Social media and the post-truth world order*, Palgrave Pivot, New York.

Crock M, Ernst C and McCallum R (2012) 'Where disability and displacement intersect: Asylum seekers and refugees with disabilities', *International Journal of Refugee Law*, 24(4): 735–764.

Crusius P (2019) *The inconvenient truth* [Due to the sensitivity of this content we do not include a link to the original document].

Curthoys A (1999) 'Expulsion, exodus and exile in white Australian historical mythology', *Journal of Australian Studies*, 23(61): 1–19.

Curthoys A and Docker J (2001) 'Introduction: Genocide: definitions, questions, settler-colonies', *Aboriginal History*, 25: 1–15.

d'Ancona, M (2017) *Post-truth: the new war on truth and how to fight back*, Random House, New York.

Daniels J (2009) *Cyber racism: white supremacy online and the new attack on civil rights*, Rowman & Littlefield Publishers, Washington.

De Gibert, O, Perez N, García-Pablos A and Cuadros M (2018) 'Hate speech dataset from a white supremacy forum' *arXiv preprint arXiv*. https://arxiv.org/abs/1809.04444

De Koster W and Houtman D (2008) '"STORMFRONT IS LIKE A SECOND HOME TO ME": on virtual community formation by right-wing extremists', *Information, Communication & Society*, 11(8): 1155–1176.

Dearden L (2019) 'Facebook allows far-right group Britain First to set up new pages and buy adverts despite vow to combat extremism', *Independent*, accessed 10 November 2022. https://www.independent.co.uk/tech/facebook-britain-first-page-far-right-advert-paul-golding-adverts-a8828386.html

Denheld B (2011) 'The politics of ned', *Agora*, 46(3): 20–23.

Di Sabato B and Hughes B (2022) 'Back to the future? The Tradwives Movement and the new forms of conservative consensus building', *Anglistica AION: An Interdisciplinary Journal*, 24(1): 25–40.

Droogan J, Waldek L, Ballsun-Stanton B and Hutchinson J (2022) 'Mapping a social media ecosystem: outlinking on Gab & Twitter amongst the Australian far-right milieu'. https://researchers.mq.edu.au/en/publications/mapping-a-social-media-ecosystem-outlinking-on-gab-amp-twitter-am

Dugin A (2012) *The fourth political theory*, Arktos, Budapest.

Dyett J and Thomas C (2019) 'Overpopulation discourse: patriarchy, racism, and the specter of ecofascism', *Perspectives on Global Development and Technology*, 18(1–2): 205–224.

Eatwell R (1992) 'Towards a new model of generic fascism', *Journal of Theoretical Politics*, 4(2): 161–194.

Eatwell R (1996) 'On defining the "fascist minimum": the centrality of ideology', *Journal of Political Ideologies*, 1(3): 303–319.

Ehrlich PR (1968) *The population bomb*, Ballantine Books, New York.

El-Khoury C (22 November 2015) 'Anti-muslim extremists: how far will they go?', *ABC: RN*, accessed 1 June 2022. https://www.abc.net.au/radionational/programs/backgroundbriefing/anti-muslim-extremists-how-far-will-they-go/6954442

Every-Palmer S, Cunningham R, Jenkins M and Bell E (2021) 'The Christchurch mosque shooting, the media, and subsequent gun control reform in New Zealand: a descriptive analysis', *Psychiatry, Psychology and Law*, 28(2): 274–285.

Farrell-Molloy J (2022) *From blood and soil to ecogram: a thematic analysis of eco-fascist subculture on Telegram* [master's thesis], Charles University, accessed 16 November 2022. https://dspace.cuni.cz/bitstream/handle/20.500.11956/178352/120430333.pdf?sequence=1

Faye JP (2002) *Le siecle des idéologies [The century of ideologies]*, Colin Pocket, Paris.

Fighel J (2009) 'The "forest jihad"', *Studies in Conflict & Terrorism*, 32(9): 802–810.

Fitzgerald J (2007) *Big white lie: Chinese Australians in white Australia*, UNSW Press, Sydney.

Forchtner B (ed) (2019a) *The far right and the environment: politics, discourse and communication*, Routledge, London.

Forchtner B (2019b) 'Eco-fascism: justifications of terrorist violence in the Christchurch mosque shooting and the El Paso shooting', *openDemocracy*, accessed 22 January 2023. https://www.opendemocracy.net/en/countering-radical-right/eco-fascism-justifications-terrorist-violence-christchurch-mosque-shooting-and-el-paso-shooting/

Forchtner B (2019c) 'Climate change and the far right', *Wiley Interdisciplinary Reviews: Climate Change*, 10(5): e604.

Forchtner B, Kroneder A and Wetzel D (2018) 'Being skeptical? Exploring far-right climate-change communication in Germany', *Environmental Communication*, 12(5): 589–604.

Furniss E (2005) 'Imagining the frontier: Comparative perspectives from Canada and Australia', in Rose DB and Davis R (eds) *Dislocating the frontier: Essaying the mystique of the outback*, ANU Press, Canberra.

Galbreath DJ and Auers D (2009) 'Green, black and brown: uncovering Latvia's environmental politics', *Journal of Baltic Studies*, 40(3): 333–348.

Gaynor A (2012) 'Antipodean eco-nazis? The organic gardening and farming movement and far-right ecology in postwar Australia', *Australian Historical Studies*, 43(2): 253–269.

Geller P (2019) 'Muslim migrant arrested in Austria following series of arson attacks', *Geller Report*, accessed 18 January 2023. https://gellerreport.com/2019/06/muslim-migrant-arson-jihad.html/

Gerard A and Weber L (2019) '"Humanitarian borderwork": identifying tensions between humanitarianism and securitization for government contracted NGOs working with adult and unaccompanied minor asylum seekers in Australia', *Theoretical Criminology*, 23(2): 266–285.

Gerlitz C and Helmond A (2013) 'The like economy: social buttons and the data-intensive web', *New Media & Society*, 15(8): 1348–1365.

Giroux H (2020) 'We must overcome our atomization to beat back neoliberal fascism', *Praxis Educativa*, 24(1): 5–16.

Global Terrorism Database (GTD) (2022) *Start: National consortium for the study of terrorism and responses to terrorism*, Global Terrorism Database, accessed 19 November 2022. https://www.start.umd.edu/research-projects/global-terrorism-database-gtd

Greene VS (2019) '"Deplorable" satire: alt-right memes, white genocide tweets, and redpilling normies', *Studies in American Humor*, 5(1): 31–69.

Griffin R (1991) *The nature of fascism*, Routledge, London.

Griffin R (2003) 'From slime mould to rhizome: an introduction to the groupuscular right', *Patterns of Prejudice*, 37(1): 27–50.

Griffiths PG (2006) *The making of White Australia: ruling class agendas 1876–1888* [PhD thesis], The Australian National University, accessed 16 November 2022. https://openresearch-repository.anu.edu.au/bitstream/1885/47107/5/01front.pdf.

Guerin C, Davey J and Fisher TJ (2021) *A snapshot of far-right activity on Gab in Australia*, Centre for Resilient and Inclusive Societies, Deakin University, Melbourne.

Hage G (2012) *White nation: fantasies of white supremacy in a multicultural society*, Routledge, London.

Hanebrink P (2018) *A specter haunting Europe: the myth of Judeo-Bolshevism*, Harvard University Press, Cambridge.

Hardin G (1974) 'Living on a lifeboat', *BioScience*, 24(10): 561–568.

Hartmann B (2010) 'Rethinking climate refugees and climate conflict: rhetoric, reality and the politics of policy discourse', *Journal of International Development: The Journal of the Development Studies Association*, 22(2): 233–246.

Hartzell SL (2020) 'Whiteness feels good here: interrogating white nationalist rhetoric on Stormfront', *Communication and Critical/Cultural Studies*, 17(2): 129–148.

Harwell D and Oremus W (2022) 'Only 22 saw the Buffalo shooting live: millions have seen it since', *The Washington Post*, accessed 10 November 2022. https://www.washingtonpost.com/technology/2022/05/16/buffalo-shooting-live-stream/

Harwood M (2021) 'Living death: imagined history and the Tarrant manifesto', *Emotions: History, Culture, Society*, 5(1): 25–50.

Hernandez Aguilar LM (2023) 'Memeing a conspiracy theory: on the biopolitical compression of the great replacement conspiracy theories', *Ethnography*. https://doi.org/10.1177/14661381221146983

Hickel J, O'Neill DW, Fanning AL and Zoomkawala H (2022) 'National responsibility for ecological breakdown: a fair-shares assessment of resource use, 1970–2017', *The Lancet Planetary Health*, 6(4): e342–e349.

Hoft K (2020) 'Muslim teen accused of starting Aussie grass fire laughs as he leaves court on Tuesday', *Gateway Pundit*, accessed 25 November 2022. https://www.thegatewaypundit.com/2020/01/muslim-teen-accused-of-starting-aussie-grass-fire-laughs-as-he-leaves-court-on-tuesday/

Hughes B, Jones D and Amarasingam A (2022) 'Ecofascism: an examination of the far-right/ecology nexus in the online space', *Terrorism and Political Violence*, 34(5): 997–1023.

Humphries A (2022) 'Australia's Black Summer bushfires were catastrophic enough: now scientists say they caused a 'deep, long-lived' hole in the ozone layer', *ABC News*, accessed 8 December 2022. https://www.abc.net.au/news/2022-08-26/black-summer-bushfires-caused-ozone-hole/101376644

Intergovernmental Panel on Climate Change (IPCC) (2023) *AR6 synthesis report: Climate change 2023*, Intergovernmental Panel on Climate Change, accessed 19 March 2023. https://www.ipcc.ch/report/sixth-assessment-report-cycle/

Jackson P (2020) 'Pioneers of world wide web fascism: the British extreme right and web 1.0.', in Littler M and Lee B (eds) *Digital Extremisms*, Palgrave Macmillan. https://doi.org/10.1007/978-3-030-30138-5.

Jasser G, McSwiney J, Pertwee E and Zannettou S (2021) '"Welcome to# GabFam": far-right virtual community on Gab', *New Media & Society*. https://doi.org/10.1177/1461444821102454

Johns A (2017) 'Flagging white nationalism "after Cronulla": from the beach to the net', *Journal of Intercultural Studies*, 38(3): 349–364.

Karbevski B (2020) *Europe's far right and its attitude towards climate change and environmentalism* [master's thesis], Universitat Autònoma de Barcelona, accessed 18 December 2022. https://ddd.uab.cat/pub/trerecpro/2020/233051/TFM_bkarbevski.pdf

Karp P (2018) 'MPs widely condemn Fraser Anning's "final solution" speech', *The Guardian*, accessed 5 November 2022. https://www.theguardian.com/australia-news/2018/aug/15/mps-widely-condemn-fraser-annings-final-solution-speech

Kimmel M (2007) 'Racism as adolescent male rite of passage: ex-Nazis in Scandinavia', *Journal of Contemporary Ethnography*, 36(2): 202–218.

Kivisto P and Rundblad G (2000) *Multiculturalism in the United States: current issues, contemporary voices*, Sage Publications, London.

Knaus C (2021) 'Daniel Andrews warns of rising antisemitism after neo-Nazi gathering in Victorian national park', *The Guardian*, accessed 18 December 2022. https://www.theguardian.com/australia-news/2021/jan/28/daniel-andrews-warns-of-rising-antisemitism-after-neo-nazi-gathering-in-victorian-national-park

Köttig M, Bitzan R and Pető A (eds) (2017) *Gender and far right politics in Europe*, Springer, Cham.

Kredo A (2019) 'ISIS tells followers to set forest fires in U.S., Europe', *The Washington Free Beacon*, accessed 15 November 2022. https://freebeacon.com/national-security/isis-tells-followers-to-set-forest-fires-in-u-s-europe/

Lazer DM, Baum MA, Benkler Y, Berinsky AJ, Greenhill KM, Menczer F, Metzger MJ, Nyhan B, Pennycook G, Rothschild D and Schudson M (2018) 'The science of fake news', *Science*, 359(6380): 1094–1096.

Lindqvist S (2007) *Terra nullius: a journey through no one's land*, Granta, London.

Loadenthal M (2022) 'Feral fascists and deep green guerrillas: infrastructural attack and accelerationist terror', *Critical Studies on Terrorism*, 15(1): 169–208.

Lockwood M (2018) 'Right-wing populism and the climate change agenda: exploring the linkages', *Environmental Politics*, 27(4): 712–732.

Lubarda B (2020) 'Beyond ecofascism? Far-right ecologism (FRE) as a framework for future inquiries', *Environmental Values*, 29(6): 713–732.

Lukács G (1952) *The destruction of reason*, Merlin Press, London.

Lyons MN (2011) 'Rising above the herd: Keith Preston's authoritarian anti-statism', *New Politics*, 7(3). https://newpol.org/rising-above-herd-keith-prestons-authoritarian-anti-statism/

Macklin G (2005) 'Co-opting the counter culture: Troy Southgate and the national revolutionary faction', *Patterns of Prejudice*, 39(3): 301–326.

Macklin G (2019) 'The Christchurch attacks: livestream terror in the viral video age', *CTC Sentinel*, 12(6): 18–29.

Macklin G (2022) 'The extreme right, climate change and terrorism', *Terrorism and Political Violence*, 34(5): 979–996.

Mann C (2018) 'The book that incited a worldwide fear of overpopulation', *Smithsonian Magazine*, accessed 23 February 2023. https://www.smithsonianmag.com/innovation/book-incited-worldwide-fear-overpopulation-180967499/

Mann M (2005) *The dark side of democracy: explaining ethnic cleansing*, Cambridge University Press, Cambridge.

Marsden S, Marino D and Ramsay G (2014) 'Forest jihad: assessing the evidence for 'popular resistance terrorism'', *Studies in Conflict & Terrorism*, 37(1): 1–17.

McAdam J (2013) 'Australia and asylum seekers', *International Journal of Refugee Law*, 25(3): 435–448.

McFadden A (2022) 'Wardens of civilisation: the political ecology of Australian far-right civilisationism', *Antipode*, 55(2): 548–573.

McGowan W (8 December 2020) 'Christchurch shooter was active with Australian far-right groups online but escaped police attention', *The Guardian*, accessed 1 June 2022. https://www.theguardian.com/world/2020/dec/08/christchurch-shooter-was-active-with-australian-far-right-groups-online-but-escaped-police-attention

McGowan M (14 May 2021) 'Australian neo-Nazi Tom Sewell charged by counter-terrorism police', *The Guardian*, accessed 1 June 2022. https://www.theguardian.com/australia-news/2021/may/14/australian-neo-nazi-tom-sewell-arrested-by-counter-terrorism-police

McIntyre L (2018) *Post-truth*, MIT Press, Cambridge.

McKenna M (1996) *The captive republic: a history of republicanism in Australia 1788–1996*, Cambridge University Press, Cambridge.

McKenna M (2002) *Looking for blackfellas' point: an Australian history of place*, UNSW Press, Sydney.

McKenna M (2012) 'Transplanted to savage shores: Indigenous Australians and British birthright in the mid nineteenth-century Australian colonies', *Journal of Colonialism and Colonial History*, 13(1). https://doi.org/10.1353/cch.2012.0009.

McKenzie N and Galloway A (23 March 2023) 'Senior neo-Nazi slips out of Australia to fight Russian army', *The Age*, accessed 1 April 2023. https://www.theage.com.au/national/senior-neo-nazi-slips-out-of-australia-to-fight-russian-army-20230322-p5cudj.html

McKenzie N and Tozer J (2021a) 'Inside racism HQ: how home-grown neo-Nazis are plotting a white revolution', *The Age*, accessed 1 June 2022. https://www.theage.com.au/national/inside-racism-hq-how-home-grown-neo-nazis-are-plotting-a-white-revolution-20210812-p58i3x.html

McKenzie N and Tozer J (2021b) '"We do not need to wait for a Christchurch": Grampians cross burning spurs call for action', *The Sydney Morning Herald*, accessed 2 November 2022. https://www.smh.com.au/politics/federal/we-do-not-need-to-wait-for-a-christchurch-grampians-cross-burning-spurs-call-for-action-20210128-p56xer.html

McNamara N (2014) 'Australian Aboriginal land management: constraints or opportunities', *James Cook University Law Review*, 21: 25.

Meadows D, Meadows D, Randers J and Behrens WW (1972) *The limits to growth: a report for the Club of Rome's project on the predicament of mankind*, Universe Books, New York.

Menadue JE (1971) *A centenary history of the Australian Natives' Association 1871–1971*, Horticultural Press, Melbourne.

Miller-Idriss C (2022) 'Tucker Carlson is the No. 1 champion of this leading far-right conspiracy', *MSNBC*, accessed 1 June 2022. https://www.msnbc.com/opinion/msnbc

-opinion/fox-s-tucker-carlson-s-biggest-racist-conspiracy-theory-n1295212?cid=sm_npd_ms_fb_ma

Mocatta G and Hawley E (2020) 'Uncovering a climate catastrophe? Media coverage of Australia's Black Summer bushfires and the revelatory extent of the climate blame frame', *M/C Journal*, 23(4). https://doi.org/10.5204/mcj.1666.

Moore A (1995) *The right road?: a history of right-wing politics in Australia*, Oxford University Press, Oxford.

Moore S and Roberts A (2022) *The rise of ecofascism: climate change and the far right*, Polity Press, Cambridge.

Moreau P (2021) 'AfD and FPÖ: rejection of immigration: The basis of a common national-populist identity?', in Tournier-Sol K and Gayte M (eds) *The faces of contemporary populism in Western Europe and the US*, Palgrave Macmillan, London.

Moses AD (2019) '"White genocide" and the ethics of public analysis', *Journal of Genocide Research*, 21(2): 201–213.

Mudde C (2019) *The far right today*, John Wiley & Sons, Hoboken.

Murphy J (2020) 'Those "climate change" fires in Australia? Many of them were set by arsonists', *The New American*, accessed 20 October 2022. https://thenewamerican.com/those-climate-change-fires-in-australia-many-of-them-were-set-by-arsonists/

Murrie L (1998) 'The Australian legend: writing Australian masculinity/writing 'Australian' masculine', *Journal of Australian Studies*, 22(56): 68–77.

National Museum Australia (2023) *Australian Natives' Association*, accessed 1 June 2022. https://www.nma.gov.au/defining-moments/resources/australian-natives-association

Nissenbaum A and Shifman L (2017) 'Internet memes as contested cultural capital: the case of 4chan's/b/board', *New Media & Society*, 19(4): 483–501.

O'Brien C (2016) *The colonial kitchen: Australia 1788–1901*, Rowman & Littlefield, Washington, DC.

Orton D (2000) 'Ecofascism: What is it?: a left biocentric analysis', *The Anarchist Library Green Web Bulletin*, accessed 20 January 2023. http://home.ca.inter.net/~greenweb/Ecofascism.html

Pantucci R and Ong K (2021) 'Persistence of right-wing extremism and terrorism in the West', *Counter Terrorist Trends and Analyses*, 13(1): 118–126.

Peucker M and Fisher TJ (2022) 'Mainstream media use for far-right mobilisation on the alt-tech online platform Gab', *Media, Culture & Society*, 45(2): 354–372.

Peucker M and Smith D (eds) (2019) *The far-right in contemporary Australia*, Springer, New York.

Pisoiu D (2015) 'Subcultural theory applied to jihadi and right-wing radicalization in Germany', *Terrorism and Political Violence*, 27(1): 9–28.

Powell A, Stratton G and Cameron R (2018) *Digital criminology: crime and justice in digital society*, Routledge, London.

Preston K, Halpin M and Maguire F (2021) 'The black pill: New technology and the male supremacy of involuntarily celibate men', *Men and Masculinities*, 24(5): 823–841.

Price M (2018a) '"It's a toxic place": how the online world of white nationalists distorts population genetics', *Science*, accessed 1 June 2022. https://www.science.org/content/article/it-s-toxic-place-how-online-world-white-nationalists-distorts-population-genetics

Price S (2018b) 'Worst drought in living memory: just don't mention climate change', *Green Left Weekly*, accessed 26 November 2022. https://www.greenleft.org.au/content/worst-drought-living-memory-just-dont-mention-climate-change

Pronczuk M and Ryckewaert K (2022) 'A racist researcher, exposed by a mass shooting', *The New York Times*, accessed 15 September 2022. https://www.nytimes.com/2022/06 /09/world/europe/michael-woodley-buffalo-shooting.html?smid=tw-share

Raspail J (1975) *The camp of the saints*, Scribner, New York.

Rhodes J, Ashe S and Valluvan S (2019) *Reframing the "left behind" race and class in post-Brexit Oldham*, University of Manchester, Manchester.

Richards I (2019) 'A dialectical approach to online propaganda: Australia's United Patriots Front, right-wing politics, and Islamic State', *Studies in Conflict & Terrorism*, 42(1–2): 43–69.

Richards I (2020) '"Sustainable development", counter-terrorism and the prevention of violent extremism: right-wing nationalism and neo-jihadism in context', in Blaustein J, Fitz-Gibbon K, Pino N W and White R (eds), *The Emerald handbook of crime, justice and sustainable development*, Emerald Publishing Limited, Bingley.

Richards I (2022) 'A philosophical and historical analysis of "Generation Identity": Fascism, online media, and the European new right', *Terrorism and Political Violence*, 34(1): 28–47.

Richards I and Jones C (2023) 'Far-right identitarianism in Australia', in Zúquete, JP and Marchi, R (eds), *Global Identitarianism*, Routledge, London.

Richards I and Wood M (2020) 'Legal and security frameworks for responding to online violent extremism: a comparison of far-right and jihadist contexts', in Ireland CA, Lewis M, Lopez A and Ireland JL (eds), *The handbook of collective violence*, Routledge. https:// doi.org/10.4324/9780429197420

Richards I, Jones C and Brinn G (2022) 'Eco-fascism online: conceptualizing far-right actors' response to climate change on Stormfront', *Studies in Conflict & Terrorism*, doi .org/10.1080/1057610X.2022.2156036.

Rimmer SH (3 May 2022) 'Clive Palmer, his money and his billboards are back. What does this mean for the 2022 federal election?' *The Conversation*, accessed 1 June 2022. https://theconversation.com/clive-palmer-his-money-and-his-billboards-are-back-what -does-this-mean-for-the-2022-federal-election-182123

Risso L (2018) 'Harvesting your soul? Cambridge Analytica and Brexit' [conference paper], Brexit Means Brexit? The Selected Proceedings of the Symposium, Akademie der Wissenschaften und der Literatur, Mainz, accessed 15 November 2022. https://www .adwmainz.de/fileadmin/user_upload/Brexit-Symposium_Online-Version.pdf#page=75.

Rose DB and Australian Heritage Commission (1996) *Nourishing terrains: Australian Aboriginal views of landscape and wilderness*, Australian Heritage Commission, Canberra.

Ross A (2017) *Against the fascist creep*, AK Press, Chico.

Ross AR and Bevensee E (2020) *Confronting the rise of eco-fascism means grappling with complex systems*, Centre for Analysis of the Radical Right (CARR), accessed 29 January 2023. https://www.paperrevolution.org/wp-content/uploads/2020/09/Ross _Bevensee_2020.3.pdf

Rueda D (2020) 'Neoecofascism: the example of the United States', *Journal for the Study of Radicalism*, 14(2): 95–125.

Schlembach R (2013) 'The "autonomous nationalists": New developments and contradictions in the German neo-Nazi movement', *Interface: A Journal for and about Social Movements*, 5(2): 295–318.

Schwartz P and Randall D (2003) *An abrupt climate change scenario and its implications for United States national security*, Jet Propulsion Laboratory Pasadena, CA, accessed 16 November 2022. https://apps.dtic.mil/sti/pdfs/ADA469325.pdf

Scott T (2019) *Facing down Facebook: Reclaiming democracy in the age of (anti) social media*, The Office of Molly Scott Cato MEP, accessed 20 January 2023. https://mollymep.org.uk/wp-content/uploads/Facebook_report_FINAL-1.pdf

Skauge-Monsen V (2022) *'"Save the bees, not refugees": far-right environmentalism meets the internet* [master's thesis], Universitetet i Oslo, accessed 16 November 2022. http://urn.nb.no/URN:NBN:no-99008.

Sklair L (2000) 'The transnational capitalist class and the discourse of globalisation', *Cambridge Review of International Affairs*, 14(1): 67–85.

Smith JK (2021) *The (re)emergence of eco-fascism: white-nationalism, sacrifice, and proto-fascism in the circulation of digital rhetoric in the ecological far-right* [PhD thesis], Baylor University, accessed 16 December 2022. https://baylor-ir.tdl.org/bitstream/handle/2104/11491/SMITH-THESIS-2021.pdf.

Sparrow J (2019) 'El Paso shooting and the rise of eco-fascism', *Eureka Street*, 29(15): 7–9.

Sparrow J (2022) ''That's what drives us to fight': Labour, wilderness and the environment in Australia', *Overland*, accessed 1 January 2023. https://overland.org.au/previous-issues/issue-246/feature-thats-what-drives-us-to-fight-labour-wilderness-and-the-environment-in-australia/

Staudenmaier P (2011a) 'Right-wing ecology in Germany: assessing the historical legacy', in Biehl J and Staudenmaier P (eds) *Ecofascism revisited: Lessons from the German experience*, New Compass Press, Porsgrunn.

Staudenmaier P (2011b) 'Fascist ecology: The "green wing" of the Nazi Party and its historical antecedents', in Biehl J and Staudenmaier P (eds) *Ecofascism revisited: Lessons from the German experience*, New Compass Press, Porsgrunn.

Staudenmaier P (2021) *Ecology contested*, New Compass Press, Porsgrunn.

Sunshine S (2008) 'Rebranding fascism: National-anarchists', *The Public Eye Magazine*, accessed 29 November 2022. https://files.libcom.org/files/Rebranding%20fascism.pdf

Tarrant B (2019) *The great replacement: towards a new society: we march ever forwards*, 8chan [Due to the sensitivity of this content, and legal action in New Zealand, we do not include a link to the original document].

Taylor B (2019) 'Alt-right ecology: ecofascism and far-right environmentalism in the United States', in Forchtner B (ed) *The far right and the environment*, Routledge, London.

Thomas C and Gosink E (2021) 'At the intersection of eco-crises, eco-anxiety, and political turbulence: a primer on twenty-first century ecofascism', *Perspectives on Global Development and Technology*, 20(1–2): 30–54.

Valencia-García LD (ed) 2020 *Far-right revisionism and the end of history: Alt/histories*, Routledge, London.

Veracini L (2007) 'Historylessness: Australia as a settler colonial collective', *Postcolonial Studies*, 10(3): 271–285.

Vosoughi S, Roy D and Aral S (2018) 'The spread of true and false news online', *Science*, 359(6380): 1146–1151.

Ward R (1965) 'Australian legend re-visited', *Australian Historical Studies*, 18(71): 171–190.

Weber D, Nasim M, Falzon L and Mitchell L (2020) '#ArsonEmergency and Australia's "Black Summer": polarisation and misinformation on social media' [conference paper], Multidisciplinary International Symposium on Disinformation in Open Online Media, Leiden, accessed 29 November 2022. https://link.springer.com/chapter/10.1007/978-3-030-61841-4_11.

WhiteRoseSociety (2023) 'We worked with @ageinvestigates to reveal that National Socialist Network ...', *[Twitter]*, accessed 1 May 2023. https://twitter.com/WhiteRoseSocAU/status/1638694583604506624

Whitsel B (2001) 'Ideological mutation and millennial belief in the American neo-Nazi movement', *Studies in Conflict & Terrorism*, 24(2): 89–106.

Wilson C (2022) 'Nostalgia, entitlement and victimhood: the synergy of white genocide and misogyny', *Terrorism and Political Violence*, 34(8): 1810–1825.

Wilson D (2020) 'For liberals, it's more fun to blame climate change than arsonists for Australian fires: the arrests show it's not so simple', *RT*, accessed 10 November 2022. https://www.rt.com/op-ed/477667-climate-change-australia-fires/

Wong J (2020) 'Revealed: QAnon Facebook groups are growing at a rapid pace around the world', *The Guardian*, accessed 12 November 2022. https://www.theguardian.com/us-news/2020/aug/11/qanon-facebook-groups-growing-conspiracy-theory

Yakushko O and De Francisco A (2022) 'The (re) emergence of eco-fascism: a history of white-nationalism and xenophobic scapegoating', in Akande A (ed) *Handbook of racism, xenophobia, and populism: all forms of discrimination in the United States and around the globe*, Springer International Publishing. https://doi.org/10.1007/978-3-031-13559-0.

Zimmerman ME (1995) 'The threat of ecofascism', *Social Theory and Practice*, 21(2): 207–238.

Zuboff S (2019) *The age of surveillance capitalism: the fight for a human future at the new frontier of power*, Profile Books, London.

Zúquete JP (2018) *The identitarians: The movement against globalism and Islam in Europe*, University of Notre Dame Press, Indiana.

4 Newsmaking on the environment

Climate change resignation and denial in the Australian media

Introduction

Australia has a long cultural and political history of far-right policies and attitudes, which frequently support unfettered extractivist and ecologically harmful economic uses of the natural environment. Its history of settler-colonialism was characterised by cultural and physical genocide against Indigenous peoples, dispossessing them of lands they had managed and cared for sustainably for tens of thousands of years (McNamara 2014). Chauvinistic racial politics that emerged in early twentieth-century Australia then reinforced white nationalism institutionally, to some extent rendering extra-institutional far-right political activism superfluous (Moore 2005; Smith 2020). At the same time, a legacy of dispossession and exploitation of Australian lands occurring through mining and other industrial activity has contributed to a present-day culture of extractivism, where exploiting natural resources without end – despite the deleterious environmental effects – is perceived as a natural and inevitable part of the Australian economy (Harvey 1996; Hickel et al. 2022).

The effects of climate change in Australia have been multifaceted, including increased frequency and intensity of heatwaves, prolonged droughts, shifting rainfall patterns and heightened risk of bushfires, with significant consequences for ecosystems, agriculture and human health (McDonald 2021; Hinkson 2022). White nationalism and radical nativism in mainstream Australian political and social life have, at the same time, been accompanied by reactionary, heterosexist and racist ideologies, which persist in relation to society and the environment and perpetuate the traditions of White Australia (Fleming and Mondon 2018).

The rise in the mainstream acceptance of far-right social attitudes in countries with European colonial histories, such as Australia, has been perpetuated through news media coverage that is characterised by 'agenda-setting power and deflection', 'euphemisation', 'trivialisation' and 'amplification' (Brown and Mondon 2021). Post-truth and the related concept of 'fake news' are then considered in this investigation as contested terms since both lack stable meaning, are attached to injurious politics, and problematically presume the existence of inclusive or representative 'real news' (Mondon and Winter 2020; Sengul 2022a). The forthcoming analysis emphasises how, despite the transformative impacts of new media environments, news media coverage of social justice issues has long been

DOI: 10.4324/9781003325437-4

conducive to the aims of political domination. This undermines the in-principle goals of public information campaigns, notwithstanding the emancipatory democratic intentions of many journalists (Ryan 2006).

This chapter considers these issues in relation to Australian newsmaking about climate change, including the act or process of making news and the mediatised political treatment of newsworthy subjects addressed therein (see Hall et al. 1975; Barak 2013; Atkins 2015). With a focus on Australia-based legacy news and political-institutional media, the analysis examines the contemporary far-right tendency to mainstream right-wing, denialist, accelerationist and resignatory climate change attitudes, via their representation in the media. It aims to demonstrate that, while different right-wing traditions of climate response are often considered distinct, in many ways they are interrelated and frequently involve historically continuous culturally and socioeconomically exclusionary elements.

The forthcoming discussion includes, first, an overview of the Australian mainstream media landscape, including the influence of concentrated corporate ownership within the sector, and reactionary social and political campaigns that gain traction partly through their representation in news media. With reference to the case study of the devastating 2019–2020 Black Summer bushfires in Eastern Australia, the next part of the chapter analyses how news and social media networks at the time became interdependent, circulating denialist conspiracy theories and misinformation about the causes of the fires. This discussion emphasises how disinformation about climate change may be received and remediated within broader news and social media ecosystems, including via the input of far-right activists, media professionals and professional politicians. The chapter next presents an account of public approaches to fossil fuel industries and GHG mitigation within conservative and right-wing political-institutional arenas in Australia, focusing on attitudes towards the gas and coal industries in the context of the country's colonial history.

The following section of the analysis provides an overview of denialist and obfuscatory climate change politics that have been promoted since the 1980s in Australia by New Right political institutions. These discourses are shaped by elected politicians, fossil fuel lobby groups and the mainstream media, reflecting persistent denialist and delay tactics designed to thwart Australia's efforts at GHG mitigation. The final section of this chapter examines memes and reductionist conspiracy narratives about climate science perpetuated in far-right alt-news media in Australia. This discussion explores the relationships between New Right lobby groups and alt-news, emphasising the ways in which far-right conspiracies developed within alt-news arenas may filter into the political mainstream, partly through mainstream media platforms.

Drawing on cases and examples from the late twentieth to the early twenty-first centuries in Australia, the analysis in this chapter highlights the points of interconnection, tension and interdependency between the multi-spectrum climate denialist right, and accelerationist and resignatory far-right political positions prominent both within and beyond Australian political institutions. As with various sections in this book, our goal in conducting this investigation is not to conflate disparate positions, but rather to illuminate the often underexamined relationships across the continuum of right-wing actors responding to the devastating effects of climate change – manifested as global heating – in Australia.

The Australian media landscape

To understand the context and implications of this study, it is essential to acknowledge how changes in the news media landscape over the past three decades have transformed international patterns of media reporting. Reporting on political and justice issues in the twenty-first century, while global and digitised, has evolved in form and content from the days of the 'broadcast era' to become more widespread, polarised and sensational (Cohen 2008). These changes have resulted, in part, from an emerging digital environment in which news is communicated to audiences on a continuous 24-hour cycle and proliferates, is remediated and is sometimes mutated through circulation via social networks (Fuchs 2017). Media coverage of sensational criminological issues, such as the far right's take on global heating, is also influenced by the globalisation and digitisation of online news, and this increasingly competitive environment has created pressures for the Australian broadcast industry to increase revenue by generating engagement through clickbait, particularly given its limited domestic audience (Meese and Hurcombe 2021; ACCC 2022; ACMA 2022).

News media industries in Australia and elsewhere have also encountered a global increase in the prevalence of far-right political activism, with journalists professing to shed light on this state of affairs through their reporting (Richards et al. 2020; Wood et al. 2022). In some cases, this involves sharing information in the public interest, but media discourses can also detract from public awareness of the complex dynamics of far-right activism and its political effects (Mondon and Winter 2020; Kemmis 2022). Far-right politicians and activists have occasionally been provided with an uncritical platform to propagandise for recruitment, presenting a sanitised version of their worldview and cultivating a personal image that is not accurate but favourable to them (Poynting and Briskman 2018; Busbridge et al. 2020; Sparrow 2021). Socially discriminatory political campaigns, sometimes referred to in the Australian context as 'culture wars' (Davis 2014), have greatly increased in prominence through these media processes, from relatively marginal extremist subcultures to the centre of Australian political life. Routinised social discrimination has thus become embedded within institutional networks and professional political communities (Davis 2019).

Symbiotic relationships between political-institutional discourses and the Australian media are underscored, for instance, in Poynting and Briskman's (2018) analysis of the normalisation of anti-Muslim prejudice in Australia during the post-2001 securitisation of Muslim communities, as part of the global war on terror. By highlighting the circulation of anti-Muslim prejudice across conservative Australian media and within parliamentary arenas, the authors demonstrate the existence of a reciprocal relationship between 'respectable' liberal discourse, which is still Islamophobic, and the extra-parliamentary agitating of extreme-right political organisations (see also Mondon and Winter 2020). Another illustrative study by Busbridge et al. (2020), reveals how the far-right conspiracy theory of 'Cultural Marxism' found its way into the popular reactionary discourse employed by Australian politicians commenting on the Safe Schools Secondary School Education Programme, whose statements were mediated online and in newspapers owned by Rupert Murdoch's News Corp. Through the media, the programme's treatment of gender and sexuality was constructed by far-right political actors as an existential threat to the cultural values and institutions of an implicitly white 'Western civilisation' (see also Maher et al. 2021).

This chapter correspondingly explores how Australian broadcast and print media coverage of diverse right-wing responses to climate change has connected climate politics to other forms of social prejudice, through concerted culture war campaigns that often incorporate multifaceted reactionary tropes. Although news media representations of environmental issues sometimes focus on the words or actions of far-right actors themselves, at other times, the behaviours and attitudes conveyed in media discourses can propagate reactionary, conservative and other right-wing elements (Bromfield et al. 2021). When the media actively promotes far-right narratives, this occurs due to the exclusivist social and economic interests promoted through climate change denialism influencing editorialism (Dunlap and McCright 2011) and to certain media, and the voices amplified therein, disparagingly associating progressive environmental activism with other social justice campaigns (Busbridge et al. 2020). Reactionary editorialising in sensational newsmaking on climate change has been particularly evident within Australian media, in part a result of the highly concentrated ownership of major news media firms as well as the personal financial interests of corporate news media owners in preventing meaningful action on climate change (McKnight 2010; Stutzer et al. 2021).

To briefly highlight the Australian news media landscape, in 2012, the International Media Concentration Research Project revealed that the consolidation of control over the country's newspapers was among the highest in the world, with the top four leading firms controlling 99% of newspaper circulation (Mueller 2012). In an independent inquiry into the country's broadcast news media sector held in the same year, Ray Finkelstein QC revealed that only three owners controlled Australia's 11 titles in the metropolitan and national segments of the daily press, while there were 444 submissions to the inquiry that referred to the concentration of Australian media as a problem. Concerns featured in Finkelstein's (2012, 280) review included a lack of diversity in the opinions that are given voice, the possibility that a handful of people (media owners or journalists) will unduly influence public opinion, and the risk of a 'decline in standards due to the absence of effective competition'.

Despite multiple sources recommending that Australia's news media concentration be reduced and that greater independent oversight be introduced in the broadcast sector, the 2017 Broadcasting Legislation Amendment (Broadcasting Reform) Act removed restrictions on media ownership and control previously established, including the 75% and 'two out of three' rules. These rules had stipulated that the population of the licence area controlled by a media company or individual could not exceed 75% of the total Australian population, while a company or individual was limited to ownership of two out of three radio, television or newspaper platforms. The change came just a year after a 2016 study found that only China and Egypt had less diverse media ownership (Papandrea and Tiffen 2016). It also followed recent speculation that, should the changes occur, regional newspapers would likely be subsumed by major metropolitan outlets, while large independent sources would merge; the latter prediction already having materialised in the controversial 2018 merger of Nine Entertainment and Fairfax Media (Dwyer 2016).

The majority of national and capital city daily newspapers in Australia are now owned by Rupert Murdoch's News Corp or Kerry Packer's Nine Network, with the exceptions being *The West Australian* (owned by Kerry Stokes's Seven West Media) and *The Canberra Times* (owned by Australian Community Media). Additionally, audiovisual media outlets exhibit a similar concentration, as the ownership of television and radio stations that dominate market share is limited to a handful of families and corporations (ACMA 2022). Although the 2017 changes in the Australian media law aimed to protect domestic corporations from competition from alternative, online and international news sources, these modifications have contributed to the growth of disinformation and misinformation about climate change. Murdoch's News Corp, which dominates the broadcast market, has historically portrayed decarbonisation as prohibitively expensive, incited fear regarding potential job losses associated with the Australian economy's transition away from fossil fuel industries, and fostered distrust about scientific evidence on climate change (McKnight 2010; Hinkson 2022). Reactionary and conservative messaging, including limited discussion of social welfare issues or a critical bias when they do feature, has also long been prevalent in News Corp media, particularly in Australia within the sole national broadsheet, *The Australian* (Iliadis et al. 2020).

In 2021, News Corp's climate change denialism notably subsided in the wake of public outrage over disinformation spread through Murdoch newspapers and the 24-hour Sky News Australia channel during the 2019 and 2020 Black Summer bushfires. In partial response to this outrage, News Corp disseminated a press release through *The New York Times* and the *Sydney Morning Herald*, announcing that Murdoch-owned outlets would advocate for net-zero emissions from the world's major economies by 2050 (Mocatta 2021). However, numerous public commentators observed that after this announcement, a significant proportion of climate science denial and mockery of climate action continued to come from high-profile commentators and bloggers associated with News Corp. These commentators include the *Herald Sun*'s Andrew Bolt, who describes global heating as a 'cult of the elites'; and the *Daily Telegraph*'s Tim Blair, who claims the climate emergency is 'bogus' – both of whom have maintained over a decade of ridicule and antagonism towards climate science (Readfearn 2021). In 2021, Graham Lloyd of *The Australian* also criticised a UN recommendation 'roadmap' for phasing out coal industries, while Sky News Australia hosts Paul Murray and Peta Credlin have questioned why Australia was being 'singled out', with Murray suggesting that Australia should withdraw from the

UN instead of succumb to international pressure (Readfearn 2021). This pattern of changing convictions also has historical precedent, as Murdoch previously signalled a shift in his outlets' stance on the issue in 2006, prior to the introduction of the company's own sustainability campaign called '1 Degree' (Readfearn 2021).

Black Summer bushfires

Climate change denial and obfuscation reached new heights in the Australian media coverage of the 2019–2020 Black Summer bushfires. These fires swept across vast areas of the country's east coast, covering an estimated 24 million hectares and directly claiming the lives of at least 33 people (Binskin et al. 2020; Humphries 2022). The event also resulted in the deaths of more than three billion animals (AIDR 2023), with analyses estimating that more than 450 additional human lives were lost due to long-term exposure to fire and smoke (Humphries 2022; Davey 2023). Up to AUD7 billion in damage from GHG emissions was incurred, with the agricultural sector losing an estimated AUD4–5 billion (University of Sydney 2021). The fires were in fact so extensive that they became a significant source of carbon emissions themselves (Bowman et al. 2021). The ideological background to Black Summer included, from 2016 to 2021, only 29% of Australian news articles on bushfires linking them to climate change, with the divide in recognising the impact of climate change on the growing frequency and intensity of bushfires being particularly stark between News Corp and non-News Corp publications (Linnenluecke and Marrone 2021).

The relatively sparse recent historical coverage of climate change as a critical factor contributing to bushfires in the Australian media sector can be said to reflect both the dominance of News Corp media in Australia and the severity of its climate change denialism (Linnenluecke and Marrone 2021). This denialism was pronounced, despite the fact that climate scientists have consistently confirmed that the 2019–2020 Australian megafires were driven, at least in part, by anthropogenic climate change (Van Oldenborgh et al. 2021). For example, although in November 2019, Rupert Murdoch declared that there were 'no climate deniers' in his company, the day after that declaration, *The Australian* published a column from mining figure and geologist Ian Plimer claiming that 'it has never been shown that human emissions of carbon dioxide drive global warming' (Readfearn 2021). In a systematic 2020 review of news media coverage of climate change in 8,612 articles across the News Corp Australia publications during the Black Summer period, including *The Australian*, the *Herald Sun*, *The Daily Telegraph* and *The Courier-Mail*, 45% were revealed to be sceptical about climate change, while almost two-thirds of opinion articles were sceptical (Bacon and Jegan 2020). The report's authors argue that, despite Murdoch's claims, 'News Corp continues to pursue its practice, honed over many years, of producing large amounts of content that derides and undermines climate science, climate scientists and climate policy change advocates' (Bacon and Jegan 2020).[1]

Recognition of the anthropogenic causes of the fires was reflected in the coverage of non-Murdoch Australian and international media, with more openly 'liberal' outlets such as *The Guardian* routinely connecting patterns of global heating to the natural disaster (Mocatta and Hawley 2020; Zheng and Bhatt 2022). The commentary in much of the progressive international social and news media overall portrayed the fires and their propagandised treatment in the Australian news media as evidence of inadequate action to address climate change, characterising Australia's lacklustre efforts at GHG mitigation as dangerously short-sighted (Khalil 28 June 2021). Indicating public awareness of the extent of denialist attitudes in news media and political associations during the fires, in the midst of the event, former NSW Fire and Rescue Commissioner Greg Mullins made several unequivocal statements, including:

> Just a 1°C temperature rise has meant the extremes are far more extreme, and it is placing lives at risk, including firefighters … But you are not allowed to talk about climate change. Well, we are, because we know what is happening … This government fundamentally doesn't like talking about climate change.
>
> (Zhou 2019)

As Chapter 3 explored, denial about the role of climate change driving the Black Summer bushfires was also apparent in far-right online websites and social media. The narratives (and conspiracies) associated with Black Summer demonstrate how right-wing tropes and beliefs propagate through seemingly unpredictable and 'rhizomatic' channels (Powell et al. 2018). This chapter's investigation seeks to highlight how, though they may initially appear decentralised and non-hierarchical, these channels and the communication patterns they amplify are actually a feature of the coordinated media campaigns of the far right.

Across various far-right, white supremacist media platforms, the unprecedented fires were portrayed as historically typical and a normal aspect of Australian life (see e.g. Chapter 3, p. 77). Their severity was frequently attributed either to the influence of the Australian Greens political party's policy on controlled burning for managing fuel loads in forested areas or to widespread arson as promoted by conspiracy theories (Mocatta and Hawley 2020; Weber et al. 2020). Both narratives are provably false. First, the Greens party has minimal influence on controlled burning policy and, moreover, generally supports the practice. Second, while deliberate bush arson does occur during Australian fire seasons, the 2019–2020 season did not witness a significant increase in such activity, let alone, as per the claim circulated among right-wing social media actors, a coordinated arson campaign by environmental activists aimed at spurring climate change action (Weber et al. 2020). This baseless narrative originated from a misinterpretation (or deliberate misrepresentation) of publicly available police reports on arson, which were subsequently shared on social media and widely disseminated on denialist media networks, often through bots and Twitter hashtags such as #ArsonEmergency (Knaus 2020; Zappone 2020). Global mainstream news outlets also picked up this deceptive theory, which then re-emerged on social networks, reinforced by links to seemingly authoritative mainstream media articles (Weber et al. 2020).

To understand how such climate change politics are integrated into broader far-right political movements, it is useful to examine the variety of conspiracy theories that surfaced during the Black Summer fires. As explored in Chapter 3, other claims circulating in related spaces online in 2020 included that the COVID-19 pandemic was a scheme orchestrated by elites or governments for nefarious population control purposes; that climate change was a hoax, or that its symptoms were purposefully generated through 'geoengineering' or 'chemtrails'; and that 5G telecommunications services were unsafe or a means of bodily or mind control (Kong et al. 2022). Along with the Australian Greens political party, other prominent targets of such theories included the Australian Federal Government, intergovernmental organisations like the UN, and liberal climate change activists such as David Attenborough and Greta Thunberg. As Chapter 3 also explored, various discussions of Black Summer on the white nationalist forums of Stormfront Downunder alleged that 'Islamist arsonists' started the fires (Khalil 2021; see Chapter 3, p. 77).

Importantly, conspiracy discourses emerged in the context of a broader political landscape that was characterised by the participation of prominent far-right politicians, adding superficial institutional legitimacy to various conspiratorial claims. For example, Barnaby Joyce, the former leader of the Australian National Party, which was a part of the governing coalition at the time, reiterated the unfounded allegation that the fires resulted from a decrease in backburning, attributing the blame to environmental activists (Taylor 2020). Pauline Hanson, the leader of the ONP, also provided direct commentary on the bushfires during Black Summer, in which she denied the existence of anthropogenic climate change. In a national television interview, when asked if she believed that climate change had contributed to the fires, she responded:

> The climate is changing. You can't tell me that taxing people and putting up their electricity costs is going to change the climate. The climate [is] changing pure to nature itself and our relation to the sun.
>
> (Sengul 2022b, 367)

The News Corp coverage of Black Summer, together with social media campaigns of the far right and speeches by professional politicians, demonstrates how climate change denial-supporting narratives that align with the interests of right-wing news outlets may be quickly adopted and recirculated without verification of their accuracy. It also underscores the role of mainstream right-wing news outlets in promoting the wider uptake of ideological tropes disseminated by extra-institutional far-right activists. A 2020 Digital News Report international survey further revealed the social impact of these communications, manifest in the prevalence of societal climate change denialist beliefs (News and Media Research Centre, University of Canberra 2020). This research indicated that individuals in Australia are more likely to deny climate change than those in most other nations, with only the US, Sweden and Norway having a higher percentage of news consumers who either believe that climate change is not serious or express uncertainty about its severity (Jericho 2020). Further indicating the influence of News Corp on these beliefs systems, approximately 30% of online Sky News Australia readers were revealed not to consider climate change to be a serious issue, while 35% of Sky News Australia television viewers concurred with this statement (Jericho 2020).

These findings illustrate the extent and power of denial when it is societally reinforced in the face of compelling evidence: if climate change is a hoax but observable events suggest its reality, then those events must be deemed suspicious and potentially part of that hoax. This case also reveals that the various levels of a right-wing online ecosystem interact with each other in multidirectional cycles of mutual influence and reinforcement. Without mistaking their distinct components, the various strands of right-leaning thought concerning the environment must be recognised as, in some ways, creating space for one another.

Fossil fuels in Australia

Globally, critical coverage of climate change denialism in the Australian media during the Black Summer bushfires reflected both the devastation of that event and Australia's role at the forefront of reactionary climate change politics in previous decades. In this context, it is important to recognise that the country has been one of the world's largest per-capita GHG emitters and one of the largest exporters of fossil fuels – particularly coal and gas (Nyberg et al. 2013) – and that its eco-systems are particularly exposed to the threats inherent in a warming climate. In 2021, the IPCC reported that global heating at 1.2°C above pre-industrial levels (World Meteorological Organization [WMO] 2021) is already causing 'cascading, compounding and aggregate impacts on cities, settlements, infrastructure, supply chains and services due to wildfires, floods, droughts, heatwaves, storms, and sea level rise' (IPCC 2022, 1). Moreover, the CSIRO and BOM identified that in Australia, in 2020, atmospheric temperatures had warmed by 1.44°C since the national records began in 1910 (Rice et al. 2021, 10).

Contemporary attitudes in Australia towards climate change have been critically shaped by the economic and social influence of its fossil fuel industries. It is important to recognise how the influence of coal and gas on Australia's political elite dates back to the 1960s, when minerals overtook wool as the primary commodity export (Brett 2020a). While coal and iron ore were at the forefront of this shift, during this period, the mining industry's social licence was challenged by Indigenous Australians, who opposed mining on their ancestral lands, and by environmentalists concerned about the industry's harmful effects on natural habitats (Blainey 2013). In response, mining companies launched a strategic public relations campaign to semantically align their interests with the national interest, convincing Australians that their prosperity relied on the mining sector, the activities of which should therefore not be restricted (Stutzer et al. 2021). This chapter is partly an account of how the mining industry has been successful in its efforts.

For instance, in the 1980s, Australia's mining industry thwarted the Hawke Labor government's attempt to implement uniform land rights legislation that would protect sacred sites, provide rights to royalties, and grant Indigenous people the power to veto mining on their lands (Brett 2020a). In contrast to other Common Law countries, the Crown in Australia owns the minerals, so this veto would have granted Indigenous land owners more rights than freehold owners. Mining companies mounted an aggressive public campaign, arguing that Indigenous groups should not possess exclusive rights. A decade later, following the High Court's rulings in the Mabo and Wik cases[2] that determined certain forms of native title had endured beyond European settlement, mining companies once again fought to ensure that the ensuing legislation would not grant any veto power over mining activities – and they succeeded (Brett 2020a). As explored in the sections below, the influence of the mining lobby has been bolstered by the pressure exerted by New Right political groups with special access to government, which, concertedly since the 1980s and escalating in the 2000s, have sought to delay meaningful political action on climate change (Brett 2020a).

Australia's enduring economic and mining legacy has resulted in a heavy reliance on fossil fuel exports (Head et al. 2013, 1), causing the country to lag behind other developed nations in terms of climate action commitments (Tobin 2017). Although the 2013–2022 Liberal Party and National Party (LNP) coalition government in Australia claimed to meet and surpass its Paris Agreement target of reducing carbon emissions by 26–28% by 2030 (Verschuer et al. 2021, 4), this government faced numerous accusations of using 'creative accounting' to inflate its reduction figures. Some allegations even suggested that Australia might only achieve a 7% reduction (Chemnick 2020), which, if replicated by other developed states, could result in a global temperature rise of 4°C before the end of the century (Verschuer et al. 2021). Recent historical Australian federal governments' alleged commitments to addressing GHG emissions at the time of this book's writing then directly conflict with its ongoing plans to exploit gas and coal reserves in Australia (Da Rimini et al. 2021, 296).

In the financial year 2019–2020, Australia garnered AUD155.5 billion in export value from coal, gas and oil (Burke 2022), and in 2021–2022, had a AUD73 billion jump in coal export earnings from the previous year (Hannam 2022). At the time of writing, Australia is the world's largest exporter of metallurgical coal and the second-largest exporter of thermal coal (Quiggan 2022). Over recent decades, open-cut coal mines in regional centres have seen significant growth, such as BHP Mitsubishi mines in Queensland, Glencore mines in NSW and Queensland, Whitehaven mines in NSW, and Rio Tinto mines in NSW and Queensland (Connor et al. 2009; Connor 2016). According to Bacon and Nash's (2012) qualitative analysis of Australian newspaper coverage on coal, which drew on a comprehensive dataset of 170 Australian newspaper articles over a one-month period, coal mining and exports, along with coal-fired power generation, were depicted as undeniably beneficial to the economy, particularly considering Australia's avoidance of the most severe effects of the 2008 Global Financial Crisis. Coverage on coal typically excluded the views of environmentalists and critics of Australia's coal industry, with economic and environmental debates often presented separately. In the data examined, only 2 out of 170 stories considered the issues together (Bacon and Nash 2012).

Australia's exceptionally high per-capita electricity consumption is a key factor driving new investments in renewable energy in Australia, with the primary aim of meeting additional demand rather than significantly reducing the reliance on fossil fuels (Hickel et al. 2022). Australia's per-person electricity demand is surpassed only by Saudi Arabia, South Korea, the US and Canada, and consequently, the Australian Government often highlights 'watts per capita' as an indicator of its efforts to address climate change (Joshi 2021a). While Australia still burns coal for much of its electricity, a 'gas-led transition' has been promoted as a way to foster economic recovery after the COVID-19 pandemic, reduce energy costs, and ensure energy security, while advancing towards a low-carbon economy (Bambrick et al. 2021).

The gas-led transition primarily refers to a strategy that emphasises the increased use of natural gas as a temporary energy source during the transition from fossil fuels to renewable energy sources. The rationale behind this approach is that natural gas, as the cleanest-burning fossil fuel, can serve as a 'bridge fuel' to help reduce GHG emissions and air pollution while renewable energy technologies are developed and deployed (Dennis 2023). Since the plan was introduced in 2020 by the LNP Morrison administration (2018–2022), the Albanese ALP government has outlined further plans to expand natural gas production, invest in gas infrastructure and support the gas industry to achieve these goals, while Australia has already overtaken Qatar as the world's largest natural gas exporter in 2022 (Statista 2023). The initial budget of the 2023 ALP Albanese government allocated AUD1.5 billion to establish a new industrial centre in Darwin Harbour, which includes gas export facilities, and at the beginning of 2023, 113 new fossil fuel projects were in progress (Dennis 2023).

Critics argue that Australia's plans to expand its use of gas contributes to environmental degradation, climate change and local harm through fossil fuel extraction and consumption, particularly when the gas is extracted via hydraulic fracturing (Sovacool 2014). They also emphasise that this transition disproportionately benefits corporations and shareholders, exacerbating economic inequality as profits concentrate among large corporations, while local communities bear the brunt of environmental and health costs (Bambrick et al. 2021). The gas industry's excessive political influence in Australia has resulted in policies that prioritise corporate interests over environmental protection and public welfare (Moss 2020). The gas-led transition is also seen to distract from and delay the necessary transition to renewable energy sources, despite Australia's immense potential for solar and wind power development (Crowley 2021b).

Consecutive Australian governments have persistently refrained from imposing moratoriums or bans on new gas and coal mines, even though such mines would be superfluous in a world transitioning towards decarbonisation. New coal and gas projects in Australia are often rationalised through the use of carbon credit policies, bolstered by reports such as the Chubb Review, spearheaded by former Chief Scientist of Australia, Ian Chubb. This review recommends measures to strengthen and streamline the Australian carbon credit market. However, the proposal has encountered scepticism due to the serious concerns surrounding the existing scheme's integrity and effectiveness (Macintosh and Butler 2023). Critics of carbon credit markets broadly point to issues related to additionality, permanence, leakage, measurement, verification, unequal benefit distribution and market volatility (Spash 2010). They argue that determining project authenticity, the unpredictability of long-term emission reductions in forestry and land-use projects, and potential emission relocation can undermine carbon credits' effectiveness at reducing GHG emissions overall (Downie 2007; Pearse 2016). Furthermore, challenges in accurately measuring emission reductions, skewed benefits towards large corporations and wealthier countries, the vulnerability of such schemes to fraud and corruption (Martin and Walters 2013; Walters and Martin 2016), and price fluctuations foster doubts about the capability of a market-based strategy to drive meaningful climate action (Spash 2010; Dennis 2023).

In support of ongoing fossil fuel extraction, the mining lobby in Australia has vilified environmentalists and cast doubt on climate science, particularly in historical efforts to maintain public support for the coal industry given its dominance in the market in recent decades. One such example of vilification involves the activist Lock the Gate alliance, which opposes the expansion of coal seam gas mining in NSW. The Minerals Council of Australia, a powerful lobby group representing mining companies, including those in the coal sector, has launched several campaigns to discredit Lock the Gate, usually involving accusing the alliance of being extremist and spreading disinformation (Fitzgerald 2006). Other members of the fossil fuel lobby – which include the Australian Petroleum Production and Exploration Association, the Queensland Resources Council, the Chamber of Minerals and Energy of Western Australia, the New South Wales Minerals Council and the South Australian Chamber of Mines and Energy – have targeted Greenpeace Australia, accusing the environmental organisation of spreading misinformation, exaggerating the risks of fossil fuel projects and using illegal or unethical tactics in their protests (Pearse 2016). For instance, Greenpeace's opposition to the proposed Carmichael coal mine in Queensland based largely on its environmental impacts on the Great Barrier Reef was met with criticism from the coal industry and its supporters, who argued that the mine is necessary for economic growth and job creation (Jolley and Rickards 2020).

An underlying assumption held by numerous right-wing commentators in the Australian media, often expressed in support of fossil fuel mining projects, is that urban residents are indifferent to the economic and social well-being of people living in regional Australia (Davison and Brodie 2005). However, although regional inhabitants in Australia have historically been more likely to depend economically on environmentally detrimental commercial industries, such as large-scale agriculture and coal mining, they have also sometimes been the worst affected by these industries (Franks et al. 2010). Climate change and fossil fuel mining activities have significant adverse impacts on agricultural industries in Australia, particularly the reduction of arable land. Climate change leads to changes in rainfall patterns, causing more frequent and severe droughts or increased rainfall intensity, as well as rising temperatures, and bushfires that cause billions of dollars of damage to agricultural crops (University of Sydney 2021). Fossil fuel mining activities increase land-use competition while causing soil degradation through soil compaction, contamination and erosion, as well as creating water pollution and depletion (Reynolds 2010). In climate change reporting, moreover, reactionary and far-right perspectives that associate climate change politics with urban elites are occasionally given a platform without adequate exposition in the media of the broader political ideologies they represent, or the real-world situations they are responding to (Moore 1995).

Within this ideological environment, the opposition of several Australian politicians to multilateral efforts aimed at reducing emissions is once more apparent. For example, the LNP vehemently publicly resisted when the 2018 IPCC report called for a significant decrease in coal use to limit global heating to 1.5°C by 2050. Deputy Prime Minister and leader of the National Party, Michael McCormack, declared that Australia would not alter its policies based on 'some sort of report' (Brown and Spiegel 2019). He further emphasised that the Liberals and Nationals in government supported small businesses, industries, farms and coal mining, with coal set to remain a part of Australia's energy mix for more than a decade (Karp 2018; Brown and Spiegel 2019). This statement was made, moreover, despite the fact that, when asked, McCormack admitted that he could not name 'a single, big policy area where the Nats have sided with the interests of farmers over the interests of miners when they come into conflict' (Brett 2020b). Climate change conspiracy theories have been even more starkly endorsed in other political settings. For example, LNP senator Gerard Rennick accused the BOM of manipulating weather records to fit the 'global warming agenda', while Pauline Hanson of the ONP has consistently denied the human-induced causes of global heating, attributing the changes to the same climate fluctuations that led to the extinction of the dinosaurs (SBS 2019).

In cases where climate change denialism was not as explicit, influential Australian politicians have nevertheless demonstrated a callous indifference to the disproportionate impact of climate change on developing countries, as, for example, when Peter Dutton, the post-2022 leader of the LNP opposition party and former Immigration Minister, was captured on microphone making a tasteless joke to former Prime Minister Tony Abbott and future Prime Minister Scott Morrison. During a visit by a government delegation to Papua New Guinea, when Abbott mentioned a delayed meeting in Port Moresby, as Papua New Guinea was running on 'Cape York time', Dutton responded, 'Time doesn't mean anything when you are, you know, about to have water lapping at your door', eliciting laughter from Abbott (Wahlquist 2015). Similarly, during a parliamentary session, the then current Prime Minister Scott Morrison waved a lump of coal, declaring that it was nothing to fear (Paul 2019). This act was widely perceived as a clear indication of the government's support for the coal industry and a lack of concern for the pressing need to transition to renewable energy sources, or for the welfare of those vulnerable to climate change impacts (Teaiwa 2019).

Despite the expressed attitudes towards climate change of politicians in government, environmental activism in Australia has been vocal about the adverse effects of fossil fuels on local communities and ecosystems for many years (Connor et al. 2009). Open-cut mines are well recognised as substantially impacting water resources, soil quality and biodiversity (van der Plank et al. 2016), while coal dust and other pollutants can cause respiratory problems and other health issues for nearby residents. Mining dust can also contain harmful substances such as heavy metals and silica, leading to respiratory problems such as asthma and lung cancer (Jenkins et al. 2013). Water pollution from mining operations can contaminate local water sources, resulting in health problems such as skin irritation, gastrointestinal issues and even neurological damage (Donoghue 2004; Morrice and Colagiuri 2013). As a 2021 Climate Council report highlighted, natural gas extraction and its use in the home can also have negative health impacts, particularly on vulnerable people such as children. In addition to its polluting nature as a fossil fuel, unconventional gas development, including fracking, involves hazardous substances that can cause serious diseases such as cancer, while the extraction of coal seam gas or shale gas with or without fracking poses additional risks, including to communities living close to gas wells (Bambrick et al. 2021).

Besides direct effects, fossil fuel mining can also produce secondary impacts on the health of individuals living near the mines. Disruptions in social and economic structures stemming from mining activities can exacerbate mental health issues, such as stress, anxiety and depression (Hossain et al. 2013). Moreover, when people, including from Indigenous communities, are compelled to leave their homes and lands due to mining operations, they may experience a loss of cultural connection and a diminished sense of belonging. This, in turn, can result in impaired emotional well-being (Albrecht 2005; Bond and Kelly 2021).

There has thus been a growing awareness of the adverse impacts of fossil fuel mining, exportation and usage in Australia, including greater acknowledgement of the fact that current government policies and actions are inadequate to tackle the climate crisis (Mortoja and Yigitcanlar 2021). Despite this change in public sentiment, however, right-wing politicians, media and activists persist in obscuring and denying the human-induced causes of climate change and disregarding its detrimental effects (Peucker and Smith 2019; Fenton-Smith 2020; Dwyer et al. 2021). The following discussion section explores how a primary strategy used by industry groups and other stakeholders aimed at delaying and diverting attention from emissions reduction in Australia has been the establishing of lobbying organisations and advocacy groups. The efforts of these groups provide an ideological backdrop and veneer of intellectual credibility supporting climate change denialism at scale in Australia.

The New Right and climate change

In the late 1970s and early 1980s, a New Right economic movement emerged in Australia, driven by interest groups, religious institutions, rural advocates and businesspeople (Moore 1995). Its broad-ranging concern was the alleged collusion between unions and ALP governments to create worker-friendly industrial conditions, which was believed to impede national economic growth and weaken Australia's global competitiveness (Moore 1995). Some ALP ministers, though, were also involved in the implementation of neoliberal New Right agendas, such as the Hawke administration (1983–1991). Some figures such as Peter Walsh, the former ALP minister and founder of the climate change lobbying firm, the Lavoisier Group (Lavoisier), also sought to amplify the politics of a regional–metropolitan divide traditionally exploited by far-right organisations in the country in the service of neoliberal economic agendas (Moore 1995).

In Australia, the term 'New Right' has maintained a relatively stable meaning over time, unlike its varying interpretations in the US and Europe. Usually associated with political tendencies within the LNP coalition of political parties that emerged in the 1970s and 1980s, the term typically refers to the growing influence or mainstream acceptance of social conservatism alongside economic neoliberalism. In Australia, economic neoliberalism is occasionally called 'economic rationalism', referring to austerity-focused approaches to welfare spending and public service provision, as well as significant deregulatory and pro-privatisation policy positions (Archer 2009).

The rise of the New Right in Australia is occasionally connected to the development of a 'political class' consisting of career politicians and their supporters, who often lack solid ties to the broader working population they purport to represent (Kelly 2019). In this context, many socially conservative LNP ministers have resisted progressive policies, such as legalising same-sex marriage (despite a 2016 referendum showing majority support among Australian voters) and humane foreign and immigration policies (Silverstein 2023). In recent years, LNP cabinet ministers in particular have consistently rejected and denied climate science while backing fossil fuel industries (Stone 1991).

Prominent figures within the Australian New Right in the 1980s and 1990s included the National Farmers Federation's Ian McLachlan, mining executives Hugh Morgan and Charles Copeman and 'corporate cowboys' like Alan Bond and Christopher Skase (Moore 1995). During the same period, Ray Evans, another early leader of the Australian New Right established several single-issue political advocacy groups pursuing radical neoliberal objectives (Kelly 2019). Through organisations such as the HR Nicholls Society, the Samuel Griffith Society, Lavoisier and the Bennelong Society, Evans and other New Right proponents aimed to shape in conservative and reactionary directions government policy and public opinion on climate change, constitutional matters, industrial relations and Indigenous issues (Kelly 2019). New Right think-tanks like the Institute of Public Affairs (IPA), the Centre for Independent Studies, the Tasman Institute and The Sydney Institute under Gerard Henderson have supported the lobbying efforts of such organisations (Moore 1995). At the same time, Australian mining industry executives, including former Western Mining Corporation CEO Hugh Morgan and former Peko-Wallsend Director and CEO Charles Copeman, also played crucial roles in promoting denialist agendas (Moore 1995, 127).

New Right lobbying significantly escalated through the efforts of Lavoisier, founded in 2000 by former ALP minister Peter Walsh and Ray Evans as a direct response to emergent pressure for the LNP Howard government to tackle climate change (Kelly 2019). In these efforts, Lavoisier partnered with the IPA, the Samuel

Griffith Society, and foreign policy research centres opposing the Kyoto Protocol, such as the Asia-Pacific Economic Cooperation Study Centre – previously based at Monash University (1997–2009), and at the time of writing located at RMIT – and denialist industry groups like the Australian Industry Greenhouse Network (Kelly 2019). While Ray Evans was Secretary, Walsh served as Lavoisier's President, having previously been Minister for Resources and Energy (1983–1984) and Minister for Finance (1984–1990) in the ALP Bob Hawke government (1983–1991).[3] While Walsh had expressed extreme animosity towards anti-logging environmentalists in his political history (Moore 1995), Lavoisier was partly founded as a response to the emergence of global environmental organisations such as the Environmental Defense Fund, Friends of the Earth and Greenpeace (Kelly 2019). Its aim was also to counter the perceived threat posed by the alliance of Old Left groups, like trade unions and New Left environmentalists,[4] as demonstrated in the Builders Labourers Federation's support for green bans in 1971 (Kelly 2019, 165).[5]

The Lavoisier board consisted entirely of individuals from the mining and energy sectors, including Ian Webber, Harold Clough, Bob Foster, Bruce Kean, Peter Murray and representatives from the National Farmers Federation (Kelly 2019, 172). Without expertise in climate science, the board relied on climate denialist scientists like Bob Carter, who received a monthly stipend of USD1,667 from the US free-market think-tank The Heartland Institute; William Kininmonth, who falsely claimed to have conducted climate research at the BOM; and Ian Plimer, a Professor of Geology at the University of Melbourne who gained mining industry directorships in return for his opposition to increasing climate change awareness (Kelly 2019, 175–177). This industry entanglement reflected a global pattern of certain scientists being compensated for promoting scepticism about the reality of climate change. In the US, for instance, environmental scientist Fred Singer received apparently unsolicited funding from ExxonMobil after challenging the harmful effects of global heating, tobacco smoke, acid rain and ozone depletion (Kelly 2019, 165).

Lavoisier sought to emphasise the alleged lack of scientific certainty surrounding climate change, including whether it was happening, human-caused, harmful or even avoidable. Initially, Lavoisier's goal was to prevent Australia from ratifying the Kyoto Protocol, which called for a 6–8% reduction in GHG emissions from 1990 levels by 2012. The group's influence was particularly noticeable during the years of the Howard LNP Coalition government (1996–2007), as it helped obtain special concessions under the Kyoto Protocol, contending that Australia's dependence on fossil fuels disproportionately impacted its economy.

In 2002, on World Environment Day, the then Prime Minister John Howard unilaterally announced that Australia would not ratify the Kyoto Protocol (Kelly 2019). Lavoisier and its broader network also criticised the potential economic consequences of implementing a carbon emissions trading scheme or a 'carbon tax' in Australia, thereby influencing the immediate abolition of such a scheme and its lack of reintroduction after 2014 when Tony Abbott became Prime Minister. These accomplishments were partly facilitated by connections Evans established with American right-wing climate denialist activists and US oil and energy company representatives at a 1996 Competitive Enterprise Institute conference in Washington, and maintained since (Kelly 2019, 168). They were also enabled via the support of organisations such as the IPA, through the dissemination and engagement strategies of conferences with politicians, and publications attracting media attention, which influenced Australian politics (Kelly 2019, 185).

Significant events in Australian climate change politics include the former LNP Prime Minister John Howard requesting special concessions in the 1997 Kyoto Agreement and the subsequent LNP Abbott government's abolishing the ALP's carbon price in July 2014. After the ALP's carbon emissions trading scheme was abandoned, Malcolm Turnbull's 2016 LNP government also dismissed an emissions intensity scheme. The National Energy Guarantee was proposed in 2017 but later scrapped by Scott Morrison's LNP government. Although a new emissions reduction plan was introduced in May 2020, Australia did not set new emission reduction targets at the Glasgow COP26 in October 2021 (Crowley 2021a).

Amid these developments, New Right lobbyists provocatively claimed that environmentalists were the new communists (Kelly 2019, 165), occasionally comparing progressive climate organisations to totalitarian regimes (Kelly 2019, 173). Australia's 'climate wars' also intersected with socially discriminatory 'culture wars', while both were characterised by disinformation campaigns aimed at undermining the credibility of climate science and safeguarding the fossil fuel industries (Joshi 2021b).

In Australia, right-wing arguments about environmental issues have recently included delay tactics predicated on market-based solutions and personal responsibility for addressing GHG emissions over government intervention. Some politicians and economists suggest that implementing carbon pricing or emissions trading schemes will encourage businesses and individuals to invest in cleaner technologies and reduce their carbon footprint (Kent 2009; Walters and Martin 2016). Others emphasise technological innovation and research and development as ways to address environmental challenges without sacrificing economic growth (Mikler and Harrison 2012). These arguments may also entail support for investment in technologies such as carbon capture and storage, nuclear energy and genetically modified crops. In addition, there are those who believe that 'environmental sustainability' can be achieved through better management of natural resources like forests, fisheries and water, using private ownership and market-based solutions to encourage sustainable management practices (Sydee and Beder 2006).

Although such solutions are not a primary focus of this book, it is worth noting here that a combination of factors renders proposed 'market-based' solutions inadequate for addressing global heating. As previously mentioned, carbon pricing falls short of achieving the necessary decrease in GHG emissions to prevent catastrophic climate change (Dennis 2023). In addition, even with the full deployment of all known emission reduction technologies at their highest feasible levels, it would still be insufficient to limit global heating to 2°C, let alone

1.5°C (Rockström et al. 2017). And while individual actions such as reducing meat consumption and driving less are important, they also cannot tackle the problem's enormous scale. Structural and systemic changes are essential, including phasing out fossil fuel production, transitioning to renewable energy sources and significantly altering production and consumption patterns, including through the realisation of post-capitalist alternatives to societal organisation (see Chapter 5, p. 171, Chapter 6, p. 211).

Alt-news media

Various alt-news media outlets engaging with climate change politics have also emerged over the past two decades. In contrast to New Right institutions, these outlets primarily challenge the growing consensus on the need for climate action from an ostensibly grassroots, anti-institutional and anti-intellectual perspective, often linking climate change politics to broader far-right and extreme-right social and political agendas. Despite their varying forms and targets, some alt-media platforms in the Australian context also maintain financial and practical ties to New Right denialist think-tanks and lobby groups. Both connected and independent far-right activists who establish alt-news outlets generally incorporate mainstream denialist discourse into the climate-related social and environmental discussions on these platforms.

Alt-news media can refer to content or form, including the sources used, the voices and perspectives emphasised and the methods of action advocated in response to pressing social issues (Rauch 2007; Haas 2004; Hamilton 2000; Örnebring and Jönsson 2004; Harcup 2003; McLeod and Hertog 1992; Platon and Deuze 2003; Rauch 2003; Rauch 2016; Holt et al. 2019). They often present alternatives to mainstream portrayals of events, issues, opinions and interpretations, covering oppositional politics, radical culture and overlooked stories (Atton 2002; Couldry and Curran 2003; Downing 2000; Hamilton 2000; Makagon 2000; Traber 1985). The 'alternative production' of alt-news can then refer to structural, financial and technological processes associated with the creation and distribution of these outlets (Rauch 2015).

Alt-media outlets typically are smaller than mainstream news organisations, and they may be non-profit, or use dialogic online communication with citizen journalists or 'producers' for content creation, rather than following a traditional linear broadcast model. They also tend to leverage cost-effective technological methods for sharing information. While progressive alt-news media can serve as a 'corrective' to misleading or misrepresentative mainstream news (Holt et al. 2019; Mayerhöffer 2021), they have historically reinforced the 'fourth estate' role of the media in holding powerful institutions accountable (Downing 2008; Atton and Wickenden 2005; Fuchs 2010). Alt-news media can represent both left- and right-wing political perspectives, addressing social harms and democratising knowledge in the public interest, or promoting social harm and violence based on counterfactual news and misinformation (Haller et al. 2019; Holt 2019, 2020). The current prominence of far-right alternative media compared to earlier periods reflects the syncretic adoption of left-wing aesthetics and popular ideology by the far right, which has been explored in relation to far-right environmental politics and related ideological trends in the previous chapters (Humprecht 2019; Nilsson 2021).

Importantly, alt-news media and mainstream media exist along a continuum, with ideas and practices extending in both directions. Mainstream media narratives can emerge in alt-news outlets, propagating counter-cultural or extreme political issues, while mainstream outlets may also incorporate alternative content, such as citizen-produced news (Kenix 2011; Harcup 2005). In what follows, we present a brief overview of the right-wing and far-right alt-news media platforms in Australia that engage with climate change politics, the tenets of which have also been circulated in the mainstream media. This list is not exhaustive but offers illustrative examples beyond the cases discussed in Chapter 3 of how environmental politics are used across far-right alt-news media for ethnonationalist, radically traditionalist and other socially reactionary purposes (see p. 93). Additionally, it builds on the previous discussion of climate change politics in Australian media and political institutions in this chapter by identifying the connections between alt-news media and New Right lobbying groups and the conspiracy theories that originate in these marginal platforms.

- The Australian National Review (ANR) is a weekly national online newspaper founded in 2014 by Jamie McIntyre, a purveyor of retirement properties. The ANR is known for promoting conspiracy theories related to an impending New World Order and Communist takeover, and its politics are characterised as conspiracist, anti-vax and antisemitic. Climate change is dismissed as a 'globalist hoax' used to control people's lives (McIntyre 3 September 2022), and the site alleges that it is part of a larger plot called 'The Great Reset agenda' that aims to 'enslave humanity' under the rule of an 'unelected few who will run the world through algorithms and AI' (Wilson 28 June 2022; Timbo 29 July 2022).
- Caldron Pool, founded in 2017 by Ben Davis, an occasional cartoonist for Spectator Australia, is a Christian nationalist alt-media site. Its politics are anti-left, anti-abortion and anti-LGBTQIA+, and it claims to demonstrate the truth of Christianity over all other religions while discerning the underlying deceptions of alternate worldviews. The site denies anthropogenic climate change and argues that an increase in natural disasters is not a result of human activity but rather a natural state of affairs in a post-Original Sin world (Jeffery 27 December 2019). Its coverage has been less concerned with evidence to prosecute that argument and more concerned with the culture wars pertaining to the general climate science debate (Caldron Pool, 24 September 2019; Lampard, 24 June 2021).
- Culture War Resource (CWR) is a conservative culture war site run by Australian end-times preacher Daniel Secomb. Its primary political interests are Christian nationalist, anti-LGBTQIA+, anti-left and anti-abortion. The site dismisses climate change reporting as a product of 'leftism' and a tool of both elites who are

planning mass third-world migration and of the UN seeking greater control over global populations (CWR Staff 21 June 2019; CWR Staff 14 December 2019; CWR Staff 16 December 2019).

- The Daily Declaration is a conservative Christian news site established by Warwick and Alison Marsh. Its politics are characterised as Christian nationalist, anti-LGBTQIA+, anti-left and anti-abortion. The site dismisses the IPCC reports as fallible due to 'Original Sin' (Balogh 3 June 2021) and alleges that climate alarmism is part of a One World Government conspiracy to promote abortion to address overpopulation. The site further asserts that God would not have given us coal if he did not want us to burn it (Smith 8 May 2020), and that climate alarmism is just another expression of Cultural Marxism (Jeffery 2 March 2020).

- Eternity News is a Christian news site operated by the Bible Society of Australia, featuring coverage on local Christian politics and news related to spreading the Gospel of Christ. Although its political stance is conservative, it tends to avoid reactionary views on issues of gender and sexuality. Eternity News almost exclusively endorses mainstream climate science and recognises the urgent need to address climate change, with rare exceptions in its coverage (Robertson 5 November 2021; Eternity News 2022).

- The Good Sauce is a Christian nationalist opinion site, with editor Dave Pellowe leading the charge against COVID-19-related lockdowns. The site's position on climate change is characterised by a rejection of climate change 'theology' and scepticism towards calls for politicians to take action on climate change. The argument is made that such calls ignore the theological and logical case against taking such action (Pellowe 21 September 2019).

- MercatorNet is a conservative Catholic opinion site edited by Michael Cook and affiliated with Opus Dei. Its main political targets are bioethics (abortion and euthanasia), transgender people, the family and demography, including denial of 'overpopulation'. In recent years, the site has positioned itself as an antidote to postmodern relativism and to 'wokeism' (Cammaerts 2022). The site largely follows the papal lead on climate change, which views it as an urgent crisis that requires action, and, as a result, takes a softer stance on climate issues than other far-right producers in the Christian media space. The site's coverage of climate change typically highlights two extremes: climate sceptics, whom they associate with evangelical Christianity, the chief rival of Catholicism; and climate alarmists, who are portrayed as atheists (Stephan 7 January 2022).

- Michael Smith News is a blog run by former 2UE radio host Michael Smith, who was dismissed from his position after becoming preoccupied with and refusing 2UE management's direction not to air an interview about the 'slush-fund' scandal that implicated then ALP Prime Minister Julia Gillard. He also lost a position at 2GB after making Islamophobic remarks. The blog reflects the changing stance of the hard right of the LNP on various social and political topics. It postulates that climate change is not real and uses the issue as a reason to attack other conservative targets such as activist journalism (Michael Smith News 26 August 2019), the infiltration of 'wokeism' into the military (Michael

Smith News 16 July 2019), Big Tech censorship (Michael Smith News 24 April 2022), left-wing journalists at the ABC (Michael Smith News 20 November 2019) and 'wets' (party members or politicians who hold more moderate or centrist views) within the Liberal Party (Michael Smith News 14 October 2019).

- New Dawn is a conspiracist magazine founded by Robert Pash, a national socialist and supporter of Muammar Gaddafi, who has been criticised for publishing Holocaust-denial content (Lee 28 November 2011). The blog has an ambivalent attitude towards climate change, often suggesting that mainstream scientists are wrong about UFOs and therefore they might be wrong about climate change as well (New Dawn, n.d.). Alternatively, the magazine has also proposed that climate change could be a natural phenomenon caused by a 'sentient sun' (Schoch 2016).

- News Weekly is a conservative Catholic magazine that is operated by the National Civic Council (see Chapter 2, p. 40) and the Australian Family Association. The magazine's political leanings align with Catholic conservatism and include stances against LGBTQIA+ communities, abortion and euthanasia. The publication's position on climate change is often communicated through articles by William Kininmonth, a climate sceptic with links to New Right lobbyists and the fossil fuel industry. These articles typically argue that while the climate might be warming, it is a natural occurrence and not a cause of concern (Kininmonth 14 September 2022). News Weekly has also published articles by Queensland Senator Matt Canavan that promote nuclear energy (Canavan 27 January 2022). Additionally, the magazine has argued that renewable energy is not viable because it requires investing money in China (Westmore 31 August 2022).

- News Blaze is an alt-news source that operates as a content farm, or a website featuring large amounts of low-quality articles designed to manipulate search engine indices, and is edited by South Australian Alan Grey. Being a global content farm, Newsblaze.com covers a wide range of topics. For example, the billionaire Jewish philanthropist George Soros, who is a common target for the far right, is portrayed as both a terrorism financier and a legendary investor spreading democracy (McCormick 3 February 2011; Schwab 2 September 2017). The Australia-focused spin-off at newsblaze.com.au was previously focused on anti-lockdown conspiracy theories, although this has changed over time. The site generally takes a sceptical view of climate change, with some of its authors having ties to oil industry-funded organisations like the Committee For A Constructive Tomorrow (McCormick 14 September 2022). However, several writers acknowledge climate change as inevitable and advocate mitigation strategies such as not buying a beachfront property and instead investing in e-books about protecting one's family from climate change (Driessen 2 May 2022).

- Quadrant is a conservative magazine edited by Keith Windschuttle, a former board member of the ABC and a prominent figure in the Australian 'history wars'. Once a member of the political left, Windschuttle has shifted to conservatism and now promotes revisionist histories that have been characterised as 'white-washing' Australia's colonial past.[6] The magazine's hard-right politics is reflected in its attitude towards climate change, which varies depending on the author. For instance, Garth Paltridge, an Australian atmospheric physicist and retired climate scientist funded by the oil industry, claims that climate change is not a problem, bushfires are caused by the Greens, and children protesting for climate action are akin to child suicide bombers (Williams 13 June 2019; Thomas 20 October 2020). However, Michael Kile, a former member of the Australian mining industry, aims to sow doubt by arguing that throughout history, people have been incorrect in their conclusions about scientific matters, so this could also be the case with climate change (16 June 2019). Meanwhile, Greg Williams, a Quadrant writer and high school maths teacher, highlights what he sees as a paradox in the concerns of younger generations about climate change, given that many are still driven to school in cars by their parents (13 June 2019).

- Quillette is an online magazine website founded by Claire Lehmann, who is also a contributor to *The Australian* and the *Sydney Morning Herald*. The site is known for its contrarian and reactionary politics. Although most of the articles on the site discussing climate change are written by Michael Shellenberger, an environmentalist, he opposes the climate change movement and the use of renewables, instead of advocating for nuclear energy. It is worth noting that Shellenberger has received funding from the nuclear industry (DeSmog 2023).
- Rebel News is an alt-right media outlet with a branch in Australia, primarily run by Avi Yemini, with contributions from Claire Lehmann and Lauren Southern. The politics of the site are anti-leftist, with only occasional engagement with climate change, such as Yemini's periodic appearance counter-protesting at climate change rallies (Yemini 2022).
- The Saltbush Club is a blog associated with climate-sceptic lobby groups, featuring contributions from former Queensland Premier Campbell Newman and Senator Malcolm Roberts. The blog takes the position that climate change is caused by solar activity and it opposes efforts to curtail GHG emissions, emphasising instead the importance of 'practical' measures for climate resilience and adaptation, such as building dams and improving planning for natural disasters (Hardaker 6 February 2020; The Saltbush Club n.d.).
- The Spectator Australia, a hard-right Australian offshoot of *The Spectator*, is edited by Rowan Dean. Its articles generally express scepticism about anthropogenic climate change, with several articles on the site claiming that emission targets are the first step towards communism (The Spectator Australia, 9 July 2022).
- The Unshackled is an alt-right media site founded by Tim Wilms, who was previously a member of the Australian Proud Boys and has co-produced media with Holocaust denier Sukith Fernando. The site's politics are far-right, with articles and videos alleging that climate change is a hoax perpetrated by the left (Richards et al. 2021).
- XYZ is an extreme-right blog primarily maintained by NSN member David Hiscox (see Chapter 3, p. 73). The site alleges that climate change is a hoax

orchestrated by 'the Jews'. Although Australian climate sceptic Stephen Wells is a frequent contributor to the site, his posts primarily feature antisemitic and white supremacist content (Wells 2023).

As the list above demonstrates, far-right actors utilise alternative media to deliver news that 'informs' their audience about various social and political issues, but this (dis)information primarily serves to cultivate indignation or moral anger directed at a purportedly guilty party (Rone 2022). Those blamed are often archetypal enemies of the far and extreme right, such as Jewish people, migrants and the political left (Frischlich et al. 2020). Conspiratorial narratives are employed to fuel this indignation, with outlets like XYZ and The Unshackled extensively engaging with the Great Replacement and other white genocide conspiracy theories (Richards et al. 2021). Coupled with the far-right alt-news media's animosity towards left-wing intellectualism, academia and 'wokeism',[7] these ideological biases provide a foundation for supremacist and counterfactual exclusionary narratives concerning potential political responses to climate change.

Perhaps unsurprisingly, hostility towards those the far right targets often coexists in this media with advocacy for social actions designed to resist and counter politico-cultural progressivism, including elements of climate change action. These activities may encompass petitioning, fundraising for far-right causes, and organising violent and nonviolent protests (Rone 2022). However, as the descriptions above indicate, denialist or sceptical stances on climate change are not uniform across far-right alt-news media in Australia. Some outlets, like The Saltbush Club, support delay and distraction measures advocated by former federal government administrations, while others, such as MercatorNet, seek to position their socially conservative agenda as 'moderate' and therefore adopt climate positions that are neither 'denialist' nor 'alarmist'. Notably, politicians, fossil fuel industry representatives and individuals with institutional connections to New Right climate delay and denialist lobby groups play a significant role in contributing to the far-right alt-news media ecosystem in Australia (see Chapter 4, p. 135).

Far-right actors, including those discussed here, can utilise alt-news media to undermine and criticise the mainstream media (Figenschou and Ihlebaek 2019). For instance, the name XYZ was chosen to reflect its opposition to Australia's national broadcaster, the ABC, which it perceives as biased in favour of the political left (Richards et al. 2021). A prevalent narrative in far-right alt-news media is that the mainstream media, particularly public news, are biased, partisan and deceptive (Figenschou and Ihlebaek 2019). Journalists are often accused of leaning towards left-wing and progressive causes and of being 'politically correct', elitist, lazy, ignorant, or exhibiting a combination of these traits (Figenschou and Ihlebaek 2019).

Furthermore, despite the hostility of many far-right alt-news outlets towards the mainstream media, it is crucial to acknowledge the practical and ideological interrelationships between different media platforms. This is particularly relevant in Australia, where, as earlier sections of this chapter discussed, the obfuscation of the reality of climate change and of its severe global impacts remains widespread in media narratives. This interdependence is even more apparent now due to the emergence of 24-hour news cycles, which have altered how social news and information are processed and received. For example, alt-right news media extensively use websites to publish commentary, analysis and news content (Figenschou and Ihlebaek 2019; Mayerhöffer 2021; Rone 2022)

Content from mainstream social media platforms such as Facebook, Twitter and YouTube also appears on alternative technology (or alt-tech) media platforms like Gab and Telegram, which far-right alt-news website publishers use to promote and disseminate content (Boberg et al. 2020). Although space constraints prevent a more detailed exploration of these interrelationships, it is crucial to acknowledge how alternative social media and alt-news media are intertwined, with the former echoing and amplifying the latter (Walther and McCoy 2021). It is also essential to recognise that the broad spectrum of right-wing media coverage of climate change in Australia operates within a more extensive field of political influence. The wider political and intellectual environment that shapes responses to economic-environmental politics in Australia entails various forms of (extra-)institutional media working together.

Concluding comments

This chapter has demonstrated that within various sectors of Australian political institutions and the media, rising international awareness and concern about climate change has been met by systematic reactionary and right-wing political responses. As the combined analysis across Chapter 3 and this chapter demonstrate, these attitudes towards global heating are often linked to racist or heterosexist views tied to the Australian 'culture wars'. In this context, those wars can refer to socially prejudicial political debates on issues such as immigration, multiculturalism, marriage equality and the human rights of non-heterosexual people, and colonisation and the rights of Indigenous people (Busbridge et al. 2020; see Chapter 3, pp. 79, 86, 101).

One notable example is the Murdoch press's denial of the anthropogenic causes and impacts of climate change until their announced change of position in 2021 (Mocatta 2021), though, as demonstrated, this shift in attitude was hardly uniform or consistent. Conspiracy theories regarding arson and the alleged role of the policy positions of the Australian Greens in driving the 2019–2020 Black Summer bushfires in Australia were promoted by News Corp media, while counterfactual coverage of climate issues in Australia more broadly involved the spread of conspiratorial extreme-right narratives within new social media ecosystems (Weber et al. 2020). Theories circulating in the mainstream international media, alt-news media and social media also drew on fearmongering and prejudice about the supposed 'cultural decay' and 'degeneracy' associated with urban societies, as well as the alleged disregard city elites hold for the economic and social needs of rural and regional communities (Moore 1995).

The chapter then presented an overview of the efforts of New Right lobby groups in Australia and their international connections, emphasising the strategic political-institutional techniques employed to foster scepticism about the dangers of fossil fuels and GHG emissions through corporate-sponsored propaganda networks (Kelly 2019). This discussion explored the New Right's links to politicians and the connections between institutional lobbying and the employment by Australia's far-right alt-news media of reactionary climate narratives and the 'memeification' of far-right denialism and conspiracy (Richards et al. 2021). New Right think-tanks and interest groups financed by the energy sector were shown to strive for intellectualism and credibility to support their denial and delay tactics concerning GHG reduction. In contrast, alt-news networks and affiliated social media platforms focused more on disseminating crude disinformation about climate change while promoting ultraconservative, reactionary or ethnonationalist far-right political positions (Figenschou and Ihlebaek 2019).

The various modes of communication explored in this chapter and throughout the book should not be confused, as the extremity of rightist positions expressed varies in target and nature. However, it is important to comprehend how New Right positions can establish an ideological context and foundation for further rightist conspiracy theories, and how the New Right's advocacy for limited government may pave the way for a rise in radically traditionalist stances (Moore 1995). For example, New Right, neoliberal advocacy of internationalisation and 'small government' has been criticised by the far right for its undermining of traditionalist social roles, ethnocultural homogeneity and 'family values', while New Right discourse that shifts the political Overton window generally to the right has made more reactionary political and social movements seem less extreme (Moore 1995). As this chapter sought to show, climate change theories that incorporate broader far-right social perspectives are then often cultivated on alt-news platforms before being assimilated into broadcast and print news media (Kenix 2011; Frischlich et al. 2020). We suggest that this process, in turn, increases the adoption of far-right attitudes towards climate change across mainstream society.

This process can also be perceived as a form of information disorder known as 'information pollution', an ecological analogy in which misinformation, disinformation and malinformation blend with an obscuring, toxic mix of political viewpoints online (Wardle and Derakhshan 2017; Milner and Phillips 2021). One facet of this theory posits the greater likelihood of accepting new, occasionally non-factual information if it aligns with pre-existing beliefs. For example, within the Australian context, climate science denialism might be more easily believed when linked to the existing biases held by a specific media audience, such as far-right notions related to Cultural Marxism, progressive environmentalism or urban elites (Phillips 2019). Knowledge-consistent material may also coalesce around far-right narratives rooted in white supremacist theories related to both social demographics and the natural environment (see e.g. Davis 2014).

Deepening the nature of the analogy, information pollution theory effectively draws on the environmental justice movement. It highlights how conspiracy ideologies often initially target vulnerable social demographics, similar to how environmental hazards affect disenfranchised local communities and how global heating's most severe consequences are experienced by vulnerable communities worldwide (Phillips 2019). Ecological thinking in this context can also raise awareness of the impacts of misinformation, such as how global heating and environmental degradation can create a contagion effect, first impacting the most vulnerable populations and then influencing the destinies of other, interconnected groups (Phillips 2019; Compton et al. 2021).

The 'feedback loops' that circulate climate change disinformation (Spohr 2017), such as that seen in the cases examined above, can thus be seen as creating what has been termed an 'illusory effect'. This effect occurs when repeated exposure to counterfactual information increases its likelihood of being accepted (Phillips 2019). The effect is further intensified by the contextual influence of anti-knowledge informational environments, which have been associated with the emergence of the 'post-truth' era, where the spread of counterfactual information creates an impression that the truth is 'up for grabs' (Phillips 2019). As discussed in Chapter 3 and throughout this book, Nietzschean perspectivist influences on the far right can also be observed as combining with emerging forms of neoliberal instrumentarianism, resulting in a situation where 'truth' is significant for far-right actors only in terms of its political utility (see Richards et al. 2022; Beiner 2018; McIntyre 2018). In relation to environmental issues and more generally, these media characteristics further reinforce the utilitarian and syncretic nature of propaganda historically employed by far-right individuals and institutions (Griffin 1993).

In a somewhat counterintuitive way, as signposted at the end of Chapter 3, we can conclude that engagement with climate change politics across the Australian media spectrum demonstrates the lack of hierarchical idea transmission among the various levels of the Australian far right. As exemplified by cases like that of mass shooter Brenton Tarrant, who identified himself as an ecofascist, the process of Australian far-right radicalisation does not depend on the top-down dissemination of ideology (e.g. Lentini 2019; Campion 2020). Tarrant embraced multiple accelerationist narratives and contemporary conspiracies, such as the Great Replacement, while advocating populationist arguments that did not deny human-induced climate change but contradicted the prevalent right-wing denialist position across different levels of Australia's far-right media (see Peucker and Smith 2019; Fenton-Smith 2020; Dunlap and Brulle 2020).

The far-right media cycle in Australia can be characterised as a chaotic system of mutual influence and multidirectional ideology transmission, which is also found in contemporary political movements that have significant internet-based subcultures. Furthermore, these subcultures serve as the ideological backdrop for individual acts of political violence, sometimes referred to as 'stochastic terrorism' (Amman and Meloy 2021). However, as this book has attempted to show so far, these behaviours and the ideologies underlying them do not emerge in a cultural void. They come from psychologically addictive media platforms and historical political-ideological contexts that endorse white nationalist perspectives.

The next chapter continues this exploration of the context in which far- and extreme-right actors capitalise on climate change and other environmental issues. It investigates intergovernmental and institutional responses to what is frequently referred to as the 'environment-security-development nexus' and examines radical grassroots reactions to both global heating and the emergence of authoritarian and ethnonationalist far-right climate change positions.

Notes

1 Their research also emphasises that the half million signatures on a 2020 petition to the Australian Parliament calling for a Royal Commission into News Corp demonstrates the level of recent public awareness and concern about News Corp's conscious intention to impede effective public policy, in the area of GHG emission mitigation specifically (Bacon and Jegan 2020).

2 The Mabo and Wik cases are two landmark legal decisions in Australia that significantly impacted Indigenous land rights and the recognition of native title. These cases marked important milestones in the struggle for Indigenous rights and the acknowledgement of the historical dispossession of Aboriginal and Torres Strait Islander peoples. The Mabo case was decided by the High Court of Australia in 1992. It centred around Eddie Mabo, a Meriam man from the Torres Strait Islands, and other plaintiffs who challenged the Queensland Government's claim to the ownership of the Murray Islands (Mer) in the Torres Strait. The High Court ruled in favour of the plaintiffs, recognising the native title rights of the Meriam people. The decision overturned the doctrine of *terra nullius*, which had previously been used to justify the British colonisation of Australia under the assumption that the land was uninhabited and unclaimed. The Wik case was decided by the High Court of Australia in 1996, building upon the principles established in the Mabo case. The Wik Peoples and the Thayorre People, two Indigenous groups from the Cape York Peninsula in Queensland, claimed native title rights over pastoral leases granted by the Queensland Government. The High Court ruled that native title could coexist with pastoral leases, provided that the rights of the leaseholders were not diminished. In situations where the rights of the native title holders and the pastoral leaseholders conflicted, the rights of the leaseholders would prevail. The Wik decision was significant as it clarified the relationship between native title and other land interests, particularly pastoral leases, which cover a significant portion of Australia's landmass.

3 During his tenure, Walsh implemented neoliberal reforms that continued under the ALP Paul Keating administration (1991–1996), including corporate tax rate reductions and curbing wage growth for workers, which violated the 1982 Statement of Agreement. Raised on a farm in Western Australia, Walsh had also historically expressed animosity towards anti-logging environmentalism.

4 The Old Left was characterised by a strong emphasis on the role of the working class, industrial action, and the importance of collective bargaining in achieving social and economic change, whereas the New Left was heavily influenced by the global counterculture movement and the student protests of the 1960s. It was characterised by a more decentralised and grassroots approach to political activism, with an emphasis on participatory democracy, direct action, and the empowerment of marginalised groups.

5 The green bans were a series of labour and environmental actions that took place in Australia, mainly in Sydney, during the early 1970s. The movement was initiated by the New South Wales Builders Labourers Federation, a trade union representing construction workers. The green bans were a unique and influential form of protest that combined environmental activism with labour rights and social justice concerns. The term 'green ban' was coined when the union decided to impose a ban on construction work at a particular site for environmental or social reasons. Union members would refuse to work on projects that they deemed harmful to the environment, heritage or local communities. This strategy was highly effective, as it directly impacted the developers' ability to complete projects and forced them to reconsider their plans. The first green ban was imposed in 1971 to protect Kelly's Bush, a patch of bushland in the

Sydney suburb of Hunters Hill, from being developed into luxury housing. The successful defence of Kelly's Bush inspired further green bans, which ultimately helped protect numerous sites across Sydney, including The Rocks, Woolloomooloo and Centennial Park.

6 The term 'history wars' refers to a long-standing and contentious debate in Australia over the interpretation and representation of the country's colonial history, particularly in relation to the treatment of Aboriginal and Torres Strait Islander peoples. The history wars have involved historians, politicians and the wider public, with disputes often focusing on issues such as the extent of violence against Indigenous peoples, the motivations of European settlers and the overall impact of colonisation on Indigenous communities.

7 The far-right perspective views 'wokeism' as a disparaging label aimed at progressives who champion social justice and inclusiveness. They perceive it as an extreme form of political correctness, assaulting traditional values and limiting free speech, while fostering a culture of victimhood, suppressing open discussion and eroding meritocracy through a focus on identity politics and topics such as race, gender and sexual orientation (Cammaerts 2022).

References

Albrecht G (2005) ''Solastalgia': a new concept in health and identity', *PAN: Philosophy Activism Nature*, 3: 41–55.

Amman M and Meloy JR (2021) 'Stochastic terrorism', *Perspectives on Terrorism*, 15(5): 2–13.

Archer V (2009) 'Dole bludgers, tax payers and the New Right: constructing discourses of welfare in 1970s Australia', *Labour History*, 96: 177–190.

Atkins BD (2015) *The newsmaking criminality of American neo-Nazi groups 1991–2011: a content analysis*, Sam Houston State University, Huntsville.

Atton C (2002) *Alternative media*, Sage, London.

Atton C and Wickenden E (2005) 'Sourcing routines and representation in alternative journalism: a case study approach', *Journalism Studies*, 6(3): 347–359.

Australian Communications and Media Authority (ACMA) (2022) *Media interests snapshot*, accessed 1 June 2022. https://www.acma.gov.au/media-interests-snapshot

Australian Competition and Consumer Commission (ACCC) (2022) *News media bargaining code*. https://www.accc.gov.au/focus-areas/digital-platforms/news-media-bargaining -code/news-media-bargaining-code

Australian Institute for Disaster Resilience (AIDR) (2023) *New South Wales, July 2019– March 2020 Bushfires: Black Summer*, Australian Disaster Resilience Knowledge Hub, accessed 1 January 2023. https://knowledge.aidr.org.au/resources/black-summer -bushfires-nsw-2019-20/

Bacon W and Jegan A (2020) *Sceptical climate: lies, debates, and silences – How News Corp produces climate scepticism in Australia*. https://climate-report.wendybacon.com /part-3/

Bacon W and Nash C (2012) 'Playing the media game: the relative (in)visibility of coal industry interests in media reporting of coal as a climate change issue in Australia', *Journalism Studies*, 13(2): 243–258.

Balogh A (3 June 2021) 'How should Christians approach climate change (and other political issues)?' *The Daily Declaration*, accessed 1 June 2022. https://blog.canberradeclaration .org.au/2021/06/03/how-should-christians-approach-climate-change-and-other-political -issues/

Bambrick H, Charlesworth KE, Bradshaw S and Baxter T (2021) *Kicking the gas habit: how gas is harming our health*, Climate Council of Australia. https://www.greengr owthknowledge.org/sites/default/files/downloads/resource/Kicking%20the%20Gas %20Habit.pdf

Barak G (2013) *Media, process, and the social construction of crime: studies in newsmaking criminology*, Routledge, London.

Beiner R (2018) *Dangerous minds: Nietzsche, Heidegger, and the return of the far right*, University of Pennsylvania Press, Philadelphia.

Binskin M, Bennett A and Macintosh A (2020) *Royal commission into natural disaster arrangements: report*, Commonwealth of Australia. https://naturaldisaster.royalcommission .gov.au/system/files/2020-11/Royal%20Commission%20into%20National%20Natural %20Disaster%20Arrangements%20-%20Report%20%20%5Baccessible%5D.pdf

Blainey G (2013) *The rush that never ended: history of Australian mining*. Melbourne University Publishing, Melbourne.

Boberg S, Quandt T, Schatto-Eckrodt T and Frischlich L (2020) 'Pandemic populism: Facebook pages of alternative news media and the corona crisis – A computational content analysis', *arXiv*. https://arxiv.org/pdf/2004.02566.pdf

Bond C and Kelly L (2021) 'Returning land to country: Indigenous engagement in mined land closure and rehabilitation', *Australian Journal of Management*, 46(1): 174–192.

Bowman DMJS, Williamson GJ, Price OF, Ndalila MN and Bradstock RA (2021) 'Australian forests, megafires and the risk of dwindling carbon stocks', *Plant, Cell & Environment*, 44(2): 347–355.

Brett J (2020a) 'Coal addiction comes at huge cost', La Trobe University. https://www .latrobe.edu.au/news/articles/2020/opinion/coal-addiction-comes-at-huge-cost

Brett J (27 June 2020b) 'Forgotten farmers, mining and anti-green invective: how the Nationals became a party for coal', *The Guardian*, accessed 1 June 2022. https://www .theguardian.com/australia-news/2020/jun/27/forgotten-farmers-mining-and-anti-green -invective-how-the-nationals-became-a-party-for-coal

Broadcasting Legislation Amendment (Broadcasting Reform) Act 2017.

Bromfield N, Page A and Sengul K (2021) 'Rhetoric, culture, and climate wars: a discursive analysis of Australian political leaders' responses to the Black Summer Bushfire crisis', in Feldman O (ed) *When politicians talk: the cultural dynamics of public speaking*, Springer Nature, New York.

Brown B and Spiegel SJ (2019) 'Coal, climate justice, and the cultural politics of energy transition', *Global Environmental Politics*, 19(2): 149–168.

Brown K and Mondon A (2021) 'Populism, the media, and the mainstreaming of the far right: *The Guardian*'s coverage of populism as a case study', *Politics*, 41(3): 279–295.

Burke PJ (2022) *On the way out: Government revenues from fossil fuels in Australia*, ANU: School of Public Policy, accessed 1 June 2022. https://taxpolicy.crawford.anu.edu.au/ sites/default/files/publication/taxstudies_crawford_anu_edu_au/2022-12/complete_wp _p_burke_dec_2022.pdf

Busbridge R, Moffitt B and Thorburn J (2020) 'Cultural Marxism: far-right conspiracy theory in Australia's culture wars', *Social Identities*, 26(6): 722–738.

Caldron Pool (24 September 2019) 'Alan Jones brutally schools child climate protesters: "you're the first generation to require air conditioning and televisions in every classroom"', accessed 1 January 2022. https://caldronpool.com/alan-jones-brutally-schools-child-climate-protesters -youre-the-first-generation-to-require-air-conditioning-and-televisions-in-every-classroom/

Cammaerts B (2022) 'The abnormalisation of social justice: the "anti-woke culture war" discourse in the UK', *Discourse & Society*, 33(6): 730–743.

Campion K (2020) 'Contemporary right wing extremism in Australia', Australian Strategic Policy Institute, accessed 1 June 2022. https://www.aspi.org.au/report/counterterrorism -yearbook-2020

Canavan M (27 January 2022) 'Nuclear: a clean, green energy machine', accessed 1 January. https://ncc.org.au/newsweekly/cover-story/nuclear-a-clean-green-energy -machine/

Chemnick J (6 January 2020) 'As fires rage, Australia pushes to emit more carbon', Scientific American, accessed 1 June 2022. https://www.scientificamerican.com/article/ as-fires-rage-australia-pushes-to-emit-more-carbon/

Cohen JE (2008) *The presidency in the era of 24-hour news*, Princeton University Press, Princeton.

Compton J, van der Linden S, Cook J and Basol M (2021) 'Inoculation theory in the post-truth era: extant findings and new frontiers for contested science, misinformation, and conspiracy theories', *Social and Personality Psychology Compass*, 15(6): doi.org/10 .1111/spc3.12602

Connor LH (2016) 'Energy futures, state planning policies and coal mine contests in rural New South Wales', *Energy Policy*, 99: 233–241.

Connor L, Freeman S and Higginbotham N (2009) 'Not just a coalmine: shifting grounds of community opposition to coal mining in Southeastern Australia', *Ethnos*, 74(4): 490–513.

Couldry N and Curran J (2003) *Contesting media power: alternative media in a networked world*, Rowman & Littlefield Publishers, Washington.

Crowley K (15 October 2021a) 'Climate wars, carbon taxes and toppled leaders: the 30-year history of Australia's climate response, in brief', *The Conversation*, accessed 1 June 2022. https://theconversation.com/climate-wars-carbon-taxes-and-toppled-leaders-the -30-year-history-of-australias-climate-response-in-brief-169545

Crowley K (2021b) 'Fighting the future: the politics of climate policy failure in Australia (2015–2020)', *Wiley Interdisciplinary Reviews: Climate Change*, 12(5): e725.

CWR Staff (21 June 2019) '"Climate emergency": Ireland set to ban private cars while planning mass third world migration', *Culture War Resource*, 1 January 2022. https:// culturewarresource.com/climate-emergency-ireland-set-to-ban-private-cars-while -planning-mass-third-world-migration/

CWR Staff (14 December 2019) 'U.N. warns it cannot escape paying punitive climate "reparations"', *Culture War Resource*, accessed 1 June 2022. https://culturewarresource .com/u-n-warns-u-s-it-cannot-escape-paying-punitive-climate-reparations/

CWR Staff (16 December 2019) 'Greta Thunberg tells cheering crowd "we will put world leaders against the wall" if they do not tackle global warming', *Culture War Resource*, accessed 1 June 2022. https://culturewarresource.com/greta-thunberg-tells-cheering -crowd-we-will-put-world-leaders-against-the-wall-if-they-do-not-tackle-global -warming/

Da Rimini F, Goodman J, Swarnakar P and Ylä-Antilla T (2021) 'Climate policy networks in Australia: dynamics of failure and possibility', *Australian Journal of Politics and History*, 67(2): 295–311.

Davey M (2 January 2023) 'More than 2,400 lives will be lost to bushfires in Australia over a decade, experts predict', *The Guardian*, accessed 1 April 2023. https://www.theguardian .com/australia-news/2023/jan/02/more-than-2400-lives-will-be-lost-to-bushfires-in -australia-over-a-decade-experts-predict

Davis M (2014) 'Neoliberalism, the culture wars and public policy', in Miller C and Orchard L (eds), *Australian public policy*, Policy Press, Bristol.

Davis M (2019) 'Transnationalising the anti-public sphere: Australian anti-publics and reactionary online media', in Peucker M and Smith D (eds), *The far-right in contemporary Australia*, Springer, New York.

Davison, G and Brodie M (ed) (2005) *Struggle country: the rural ideal in twentieth century Australia*, Monash University ePress, Melbourne.

Dennis R (11 January 2023) 'As long as Australia fails to transition away from fossil fuels, its climate policy is meaningless', *The Guardian*, accessed 1 June 2022. https://www.theguardian.com/commentisfree/2023/jan/11/as-long-as-australia-fails-to-transition-away-from-fossil-fuels-its-climate-policy-is-meaningless

DeSmog (2023) 'Michael Shellenberger', accessed 1 January 2023. https://www.desmog.com/michael-shellenberger/

Donoghue AM (2004) 'Occupational health hazards in mining: an overview', *Occupational Medicine*, 54(5): 283–289.

Downie C (2007) *Carbon offsets: saviour or cop-out?*, Australia Institute.

Downing JDH (2000) *Radical media: rebellious communication and social movements*, Sage, London.

Downing J (2008) 'Social movement theories and alternative media: an evaluation and critique', *Communication, Culture & Critique*, 1(1): 40–50.

Driessen P (2 May 2022) 'Real threats to biodiversity and humanity', *News Blaze*, accessed 1 June 2022. https://newsblaze.com/issues/environment/threats-to-biodiversity_183891/

Dunlap RE and Brulle, RJ (2020) 'Sources and amplifiers of climate change denial', in Holmes DC and Richardson LM (eds) *Research handbook on communicating climate change*, Edward Elgar Publishing, Cheltenham.

Dunlap RE and McCright AM (2011) 'Organized climate change denial', in Dryzek JS, Norgaard RB and Schlosberg D (eds) *The Oxford handbook of climate change and society*, Oxford University Press, Oxford.

Dwyer T (12 December 2016) 'FactCheck: is Australia's level of media ownership concentration one of the highest in the world?', *The Conversation*, accessed 1 June 2022. https://theconversation.com/factcheckis-australias-level-of-media-ownership-concentration-one-of-the-highest-in-the-world-68437

Dwyer T, Wilding D and Koskie T (2021) 'Australia: media concentration and deteriorating conditions for investigative journalism', in Trappel J and Tomaz T (eds) *The media for democracy monitor 2021: how leading news media survive digital transformation*, University of Gothenburg, Nordicom.

Eternity News (2022) *Climate change*. https://www.eternitynews.com.au/topics/climate-change/

Fenton-Smith B (2020) 'The (re)birth of far-right populism in Australia: the appeal of Pauline Hanson's persuasive definitions', in Kranert M (ed) *Discursive approaches to populism across disciplines*, Palgrave Macmillan, London.

Figenschou TU and Ihlebaek KA (2019) 'Media criticism from the far-right: attacking from many angles', *Journalism Practice*, 13(8): 901–905.

Finkelstein R (2012) *Report of the independent inquiry into the media and media regulation, assisted by M Ricketson, Australian Government: report to the Minister for Broadband, Communications and Digital Economy*. http://www.abc.net.au/mediawatch/transcripts/1205_finkelstein.pdf

Fitzgerald J (2006) *Lobbying in Australia: you can't expect anything to change if you don't speak up*, Rosenberg Publishing Pty Limited.

Fleming A and Mondon A (2018) 'The radical right in Australia', in Rydgren J (ed) *The Oxford handbook of the radical right*, Oxford University Press, Oxford.

Franks DM, Brereton D and Moran CJ (2010) 'Managing the cumulative impacts of coal mining on regional communities and environments in Australia', *Impact Assessment and Project Appraisal*, 28(4): 299–312.

Frischlich L, Klapproth J and Brinkschulte F (2020) 'Between mainstream and alternative: co-orientation in right-wing populist alternative news media', in van Dujin M, Preuss M, Spaiser V, Takes F and Verberne S (eds) *Multidisciplinary international symposium on disinformation in open online media*, Springer, New York.

Fuchs C (2010) 'Alternative media as critical media', *European Journal of Social Theory*, 13(2): 173–192.

Fuchs C (2017) *Social media: a critical introduction*, Sage, London.

Griffin R (1993) *The nature of fascism*, Routledge, London.

Haas T (2004) 'Research note: alternative media, public journalism and the pursuit of democratization', *Journalism Studies*, 5(1): 115–122.

Hall S, Clarke J, Critcher C, Jefferson T and Roberts B (1975) *Newsmaking and crime*, Centre for Contemporary Cultural Studies, Birmingham.

Haller A, Holt K and de La Brosse R (2019) 'The 'other' alternatives: political right-wing alternative media', *Journal of Alternative and Community Media*, 4(1): 1–6.

Hamilton J (2000) 'Alternative media: conceptual difficulties, critical possibilities', *Journal of Communication Inquiry*, 24(4): 357–378.

Hannam P (15 December 2022) 'Australia's coal exporters made windfall gain of $45bn last year, report estimates', *The Guardian*, accessed 1 June 2023. https://www.theguardian .com/environment/2022/dec/15/australias-coal-exporters-made-windfall-profit-of-45bn -last-year-report-estimates

Harcup T (2003) '"The unspoken – said": the journalism of alternative media', *Journalism*, 4(3): 356–376.

Harcup T (2005) '"I'm doing this to change the world": journalism in alternative and mainstream media', *Journalism Studies*, 6(3): 361–374.

Hardaker D (6 February 2020) 'Bush logic: behind the influential group fuelling climate denialism', *Crikey*, accessed 1 January 2023. https://www.crikey.com.au/2020/02/06/ saltbush-club-part-one/

Harvey D (1996) *Justice, nature and the geography of difference*, Blackwell, London.

Head L, Adams M, McGregor H and Toole S (2013) 'Climate change and Australia', *WIREs Climate Change*, 5(2): 175–197.

Hickel J, O'Neill DW, Fanning AL and Zoomkawala H (2022) 'National responsibility for ecological breakdown: a fair-shares assessment of resource use, 1970–2017', *The Lancet Planetary Health*, 6(4): e342–e349.

Hinkson M (2022) 'Contesting rural Australia in the time of accelerating climate change', *Journal of Rural Studies*, 95: 50–57.

Holt K (2019) *Right-wing alternative media*, Routledge, London.

Holt K (2020) 'Populism and alternative media', in Krämer B and Holtz-Bacha C (eds) *Perspectives on populism and the media: avenues for research*, Nomos Verlag, Baden-Baden.

Holt K, Ustad Figenschou T and Frischlich L (2019) 'Key dimensions of alternative news media', *Digital Journalism*, 7(7): 860–869.

Hossain D, Gorman D, Chapelle B, Mann W, Saal R and Penton G (2013) 'Impact of the mining industry on the mental health of landholders and rural communities in southwest Queensland', *Australasian Psychiatry*, 21(1): 32–37.

Humphries A (2022) 'Australia's Black Summer bushfires were catastrophic enough. Now scientists say they caused a "deep, long-lived" hole in the ozone layer', *ABC News*,

accessed 8 December 2022. https://www.abc.net.au/news/2022-08-26/black-summer
-bushfires-caused-ozone-hole/101376644

Humprecht E (2019) 'Where "fake news" flourishes: a comparison across four Western democracies', *Information, Communication & Society*, 22(13): 1973–1988.

Iliadis M, Richards I and Wood MA (2020) 'Newsmaking criminology in Australia and New Zealand: results from a mixed methods study of criminologists' media engagement', *Australian & New Zealand Journal of Criminology*, 53(1): 84–101.

International Panel on Climate Change (IPCC) (2022) *Fact sheet: Australasia*, International Panel on Climate Change. https://www.ipcc.ch/report/ar6/wg2/downloads/outreach/IPCC_AR6_WGII_FactSheet_Australasia.pdf

Jeffery J (27 December 2019) 'The bushfires have nothing to do with "climate change"', *Caldron Pool*, accessed 1 January 2023. https://caldronpool.com/the-bushfires-have
-nothing-to-do-with-climate-change/

Jeffery J (2 March 2020) 'Mardi Gras, climate alarmism and Cultural Marxism', *The Daily Declaration*, accessed 1 June 2022. https://blog.canberradeclaration.org.au/2020/03/02/
mardi-gras-climate-alarmism-and-cultural-marxism/

Jenkins WD, Christian WJ, Mueller G and Robbins KT (2013) 'Population cancer risks associated with coal mining: a systematic review', *PloS One*, 8(8): doi.org/10.1371/journal.pone.0071312.

Jericho G (16 June 2020) 'Australia has a problem with climate change denial: the message just isn't getting through', *The Guardian*, accessed 1 June 2022. https://www
.theguardian.com/business/grogonomics/2020/jun/16/australians-arent-worried-about
-climate-change-the-message-just-isnt-getting-through

Jolley C and Rickards L (2020) 'Contesting coal and climate change using scale: emergent topologies in the Adani mine controversy', *Geographical Research*, 58(1): 6–23.

Joshi K (29 March 2021a) 'Australia still addicted to coal despite huge growth in wind and solar', *Renew Economy*, accessed 1 June 2022. https://reneweconomy.com.au/australia
-still-addicted-to-coal-despite-huge-growth-in-wind-and-solar/

Joshi K (30 December 2021b) 'In Australia's climate wars, delay and deception are the new denial', *The Guardian*, accessed 1 June 2022. https://www.theguardian.com/commentisfree
/2021/dec/30/in-australias-climate-wars-delay-and-deception-are-the-new-denial

Karp P (2018) 'MPs widely condemn Fraser Anning's "final solution" speech', *The Guardian*, accessed 5 November 2022. https://www.theguardian.com/australia news/
2018/aug/15/mps-widely-condemn-fraser-annings-final-solution-speech

Kelly D (2019) *Political troglodytes and economic lunatics: the hard right in Australia*, Black Inc, Melbourne.

Kemmis S (2022) 'Addressing the climate emergency: a view from the theory of practice architectures', *The Journal of Environmental Education*, 53(1): 42–53.

Kenix LJ (2011) *Alternative and mainstream media: the converging spectrum*, A&C Black, London.

Kent J (2009) 'Individualized responsibility and climate change: "if climate protection becomes everyone's responsibility, does it end up being no-one's?"', *Cosmopolitan Civil Societies: An Interdisciplinary Journal*, 1(3): 132–149.

Khalil L (2021) 'The impact of natural disasters on violent extremism', *ASPI Counterterrorism Yearbook 2021*, 107–112. https://www.jstor.org/stable/pdf/resrep31258.24.pdf

Khalil S (28 June 2021) 'Climate change: why action still ignites debate in Australia', *The Guardian*, accessed 1 January 2023. https://www.bbc.com/news/world-australia
-57606398

Kile M (16 June 2019) 'Now they want to "fix" the climate', *Quadrant*, accessed 1 January 2023. https://quadrant.org.au/opinion/doomed-planet/2019/06/now-they-want-to-fix-the-climate/

Kininmonth W (14 September 2022) 'The best of times, not the worst of times: the delusions of climate hysteria', accessed 1 January 2023. https://ncc.org.au/newsweekly/energy-science-enviro/delusions-of-climate-hysteria-global-warming/

Knaus C (8 January 2020) 'Bots and trolls spread false arson claims in Australian fires "disinformation campaign"', *The Guardian*. https://www.theguardian.com/australia-news/2020/jan/08/twitter-bots-trolls-australian-bushfires-social-media-disinformation-campaign-false-claims

Kong Q, Booth E, Bailo F, Johns A and Rizoiu M-A (2022) 'Slipping to the extreme: a mixed method to explain how extreme opinions infiltrate online discussions' [conference presentation], *Proceedings of the International AAAI Conference on Web and Social Media*.

Lampard R (24 June 2021) 'High Court ruling on Dr Peter Ridd's academic freedom case will take 2–3 months', *Cauldron Pool*, accessed 1 June 2022. https://caldronpool.com/high-court-ruling-on-dr-peter-ridds-academic-freedom-case-will-take-2-3-months/

Lee A (28 November 2011) 'Gaddafi's Australian groupies', *Australia/Israel Review*, accessed 1 January 2022. https://aijac.org.au/australia-israel-review/gaddafi-s-australian-groupies/

Lentini P (2019) 'The Australian far-right: an international comparison of fringe and conventional politics', in Peucker M and Smith D (eds) *The far-right in contemporary Australia*, Springer, New York.

Linnenluecke MK and Marrone M (2021) 'Air pollution, human health and climate change: newspaper coverage of Australian bushfires', *Environmental Research Letters*, 16(12), doi: 10.1088/1748-9326/ac3601

Macintosh A and Butler D (9 January 2023) 'Chubb review of Australia's carbon credit scheme falls short: and problems will continue to fester', *The Conversation*, accessed 1 June 2022. https://theconversation.com/chubb-review-of-australias-carbon-credit-scheme-falls-short-and-problems-will-continue-to-fester-197401

Maher H, Gunaydin E and McSwiney J (2021) 'Western civilizationism and white supremacy: the Ramsay Centre for Western Civilisation', *Patterns of Prejudice*, 55(4): 309–330.

Makagon D (2000) 'Accidents should happen: cultural disruption through alternative media', *Journal of Communication Inquiry*, 24(4): 430–447.

Martin P and Walters R (2013) 'Fraud risk and the visibility of carbon', *International Journal for Crime, Justice and Social Democracy*, 2(2): 27–42.

Mayerhöffer E (2021) 'How do Danish right-wing alternative media position themselves against the mainstream? Advancing the study of alternative media structure and content', *Journalism Studies*, 22(2): 119–136.

McCormick J (3 February 2011) 'Legendary investor George Soros comments on Egyptian solution', *News Blaze*, accessed 1 June 2022. https://newsblaze.com/world/middle-east/legendary-investor-george-soros-comments-on-egyptian-situation_17721/

McCormick J (14 September 2022) 'Mitigation and climate change: stop begging politicians and begin mitigation', *News Blaze*, accessed 1 January 2023. https://newsblaze.com/issues/environment/mitigation-and-climate-change-stop-begging-politicians_186076/

McDonald M (2021) 'After the fires? Climate change and security in Australia', *Australian Journal of Political Science*, 56(1): 1–18.

McIntyre J (3 September 2022) 'Is climate change being used by the globalists to introduce the great reset totalitarian agenda', *Australian National Review*, accessed 1 January 2023. https://www.australiannationalreview.com/state-of-affairs/is-climate-change -being-used-by-the-globalists-to-introduce-the-great-reset-totalitarian-agenda/

McIntyre L (2018) *Post-truth*, MIT Press, Cambridge.

McKnight D (2010) 'A change in the climate? The journalism of opinion at News Corporation', *Journalism*, 11(6): 693–706.

McLeod DM and Hertog JK (1992) 'The manufacture of public opinion by reporters: informal cues for public perceptions of protest groups', *Discourse & Society*, 3(3): 259–275.

McNamara N (2014) 'Australian Aboriginal land management: constraints or opportunities', *James Cook University Law Review*, 21: 25.

Meese J and Hurcombe E (2021) 'Facebook, news media and platform dependency', *New Media & Society*, 23(8): 2367–2384.

Michael Smith News (16 July 2019) 'ADF chief Angus Campbell warns about threat of invading hordes of climate change boogeymen', accessed 1 January 2023. https://www .michaelsmithnews.com/2019/07/adf-chief-angus-campbell-warns-about-threat-of -invading-hordes-of-climate-change-boogeymen.html

Michael Smith News (26 August 2019) 'The global coordination behind a week of so-called news stories on climate change', accessed 1 January 2023. https://www .michaelsmithnews.com/2019/08/the-global-coordination-behind-a-week-of-so-called -news-stories-on-climate-change.html

Michael Smith News (14 October 2019) 'Brilliant Rowan Dean on renegade climate change Libs "How dare you!"', accessed 1 January 2023. https://www.michaelsmithnews.com /2019/10/brilliant-rowan-dean-on-renegade-climate-change-libs-how-dare-you.html

Michael Smith News (20 November 2019) 'Ita tells 'em! ABC staff "climate crisis group" will not happen under Chairwoman Ita's watch', accessed 1 January 2023. https://www .michaelsmithnews.com/2019/11/ita-tells-em-abc-staff-climate-crisis-group-will-not -happen-under-chairwoman-itas-watch.html

Michael Smith News (24 April 2022) 'Twitter bans ads contrary to the IPCC 'science' on the climate', accessed 1 January 2023. https://www.michaelsmithnews.com/2022/04/ twitter-bans-ads-contrary-to-the-ipcc-science-on-the-climate.html

Mikler J and Harrison NE (2012) 'Varieties of capitalism and technological innovation for climate change mitigation', *New Political Economy*, 17(2): 179–208.

Milner RM and Phillips W (2021) *You are here: a field guide for navigating polarized speech, conspiracy theories, and our polluted media landscape*, MIT Press, Cambridge.

Mocatta G (18 October 2021) 'What's behind News Corp's new spin on climate change?', *The Conversation*, 1 June 2022. https://theconversation.com/whats-behind-news-corps -new-spin-on-climate-change-169733

Mocatta G and Hawley E (2020) 'Uncovering a climate catastrophe? Media coverage of Australia's Black Summer bushfires and the revelatory extent of the climate blame frame', *M/C Journal*, 23(4). DOI: https://doi.org/10.5204/mcj.1666.

Mondon A and Winter A (2020) *Reactionary democracy: how racism and the populist far right became mainstream*, Verso Books, London.

Moore A (1995) *The right road?: a history of right-wing politics in Australia*, Oxford University Press, Oxford.

Moore A (2005) 'Writing about the extreme right in Australia', *Labour History*, 89: 1–15.

Morrice E and Colagiuri R (2013) 'Coal mining, social injustice and health: a universal conflict of power and priorities', *Health & Place*, 19: 74–79.

Mortoja MG and Yigitcanlar T (2021) 'Public perceptions of peri-urbanism triggered climate change: survey evidence from South East Queensland, Australia', *Sustainable Cities and Society*, 75, doi.org/10.1016/j.scs.2021.103407

Moss J (2020) 'Australia: an emissions super-power', University of New South Wales. https://apo.org.au/sites/default/files/resource-files/2020-07/apo-nid306756.pdf.

Mueller H (2012) *International media concentration*, Columbia Institute for Tele-Information. http://internationalmedia.pbworks.com/w/page/20075656/FrontPage

New Dawn (n.d.) 'Science was wrong: an interview with Stanton Friedman', accessed 1 January 2023. https://www.newdawnmagazine.com/articles/science-was-wrong-an-interview-with-stanton-friedman

News and Media Research Centre, University of Canberra (2020) *Digital News Report: Australia* 2020. https://apo.org.au/node/305057

Nilsson P-E (2021) '"The new extreme right": uncivility, irony, and displacement in the French re-information sphere', *Nordicom Review*, 42(s1): 89–102.

Nyberg D, Spicer A and Wright C (2013) 'Incorporating citizens: corporate political engagement with climate change in Australia', *Organization*, 20(3): 433–453.

Örnebring H and Jönsson AM (2004) 'Tabloid journalism and the public sphere: a historical perspective on tabloid journalism', *Journalism Studies*, 5(3): 283–295.

Papandrea F and Tiffen R (2016) 'Media ownership and concentration in Australia', in Noam EM (ed) *Who owns the world's media?: media concentration and ownership around the world*, Oxford University Press, Oxford.

Paul S (2019) 'In coal we trust: Australian voters back PM Morrison's faith in fossil fuel', Reuters. http://www.burnmorecoal.com/wp-content/uploads/2019/05/In-coal-we-trust-Australian-voters-back-PM-Morrisons-faith-in-fossil-fuel-Reuters.pdf

Pearse R (2016) 'The coal question that emissions trading has not answered', *Energy Policy*, 99: 319–328.

Pellowe D (21 September 2019) 'The sins of climate change theology', *The Good Sauce*, accessed 1 June 2022. https://goodsauce.news/the-sins-of-climate-change-theology/

Peucker M and Smith D (eds) (2019) *The far-right in contemporary Australia*, Springer, New York.

Phillips W (2019) 'The toxins we carry', *Columbia Journalism Review*. https://www.cjr.org/special_report/truth-pollution-disinformation.php

Platon S and Deuze M (2003) 'Indymedia journalism: a radical way of making, selecting and sharing news?', *Journalism*, 4(3): 336–355.

Powell A, Stratton G and Cameron R (2018) *Digital criminology: crime and justice in digital society*, Routledge, London.

Poynting S and Briskman L (2018) 'Islamophobia in Australia: from far-right deplorables to respectable liberals', *Social Sciences*, 7(11): 213.

Quiggan J (6 April 2022) 'Time's up: why Australia has to quit stalling and wean itself off fossil fuels', *The Conversation*, accessed 1 June 2022. https://theconversation.com/times-up-why-australia-has-to-quit-stalling-and-wean-itself-off-fossil-fuels-180666

Rauch J (2003) 'Rooted in nations, blossoming in globalization? A cultural perspective on the content of a 'northern' mainstream and a 'southern' alternative news agency', *Journal of Communication Inquiry*, 27(1): 87–103.

Rauch J (2007) 'Activists as interpretive communities: rituals of consumption and interaction in an alternative media audience', *Media, Culture & Society*, 29(6): 994–1013.

Rauch J (2015) 'Exploring the alternative–mainstream dialectic: what "alternative media" means to a hybrid audience', *Communication, Culture & Critique*, 8(1): 124–143.

Rauch J (2016) 'Are there still alternatives? Relationships between alternative media and mainstream media in a converged environment', *Sociology Compass*, 10(9): 756–767.

Readfearn G (8 September 2021) 'Is Rupert Murdoch's News Corp Australia really shifting away from "climate denialism"?', *The Guardian*, accessed 1 June 2022. https://www .theguardian.com/media/2021/sep/08/is-rupert-murdochs-news-corp-australia-really -shifting-away-from-climate-denialism

Reynolds MP (ed) (2010) *Climate change and crop production*, Cabi, Wallingford.

Rice M, Weisbrot E, Bradshaw S, Steffen W, Hughes L, Bambrick H, Charlesworth KE, Hutley N and Upton L (2021) *Game, set, match: calling time on climate inaction*, Climate Council of Australia. https://www.climatecouncil.org.au/wp-content/uploads /2021/02/Game-Set-Match-Calling-Time-on-Climate-Inaction-Climate-Council-Sports -Report-1.pdf

Richards I, Jones C and Brinn G (2022) 'Eco-fascism online: conceptualizing far-right actors' response to climate change on stormfront', *Studies in Conflict & Terrorism*. https://doi.org/10.1080/1057610X.2022.2156036

Richards I, Rae M, Vergani M and Jones C (2021) 'Political philosophy and Australian far-right media: a critical discourse analysis of The Unshackled and XYZ', *Thesis Eleven*, 163(1): 103–130.

Richards I, Wood MA and Iliadis M (2020) 'Newsmaking criminology in the twenty-first century: an analysis of criminologists' news media engagement in seven countries', *Current Issues in Criminal Justice*, 32(2): 125–145.

Robertson D (5 November 2021) 'Is the climate change debate over?', *Eternity News*, accessed 1 June 2022. https://www.eternitynews.com.au/opinion/is-the-climate-change -debate-over/

Rockström J, Gaffney O, Rogelj J, Meinshausen M, Nakicenovic N and Schellnhuber HJ (2017) 'A roadmap for rapid decarbonization', *Science*, 355(6331): 1269–1271.

Rone J (2022) 'Far right alternative news media as "indignation mobilization mechanisms": how the far right opposed the Global Compact for Migration', *Information, Communication & Society*, 25(9): 1333–1350.

Ryan M (2006) 'Public(s), politicians and punishment', *Criminal Justice Matters*, 64(1): 14–15.

SBS (23 April 2019) 'Hanson denies humans behind climate change, blames "feamongering"', *SBS*, accessed 1 June 2022. https://www.sbs.com.au/news/article/hanson-denies-humans -behind-climate-change-blames-fearmongering/2y23ihwxy

Schoch R (December 2016) 'Is our sun conscious?', accessed 1 June 2022. https://www .newdawnmagazine.com/articles/is-our-sun-conscious

Schwab D (2 September 2017) 'George Soros the terrorist financier', *News Blaze*, accessed 1 June 2022. https://newsblaze.com/usnews/politics/george-soros-is-a-terrorist_84660/

Sengul K (2022a) '"It's OK to be white": the discursive construction of victimhood, "anti-white racism" and calculated ambivalence in Australia', *Critical Discourse Studies*, 19(6): 593–609.

Sengul K (2022b) 'The role of political interviews in mainstreaming and normalizing the far-right: a view from Australia', in Feldman O (ed), *Adversarial political interviewing: worldwide perspectives during polarized times*, Springer Nature Singapore, Singapore.

Silverstein J (2023) *Cruel care: a history of children at our borders*, Monash University Press, Melbourne.

Smith C (8 May 2020) 'Climate: a biblical view vs the world's view', *The Daily Declaration*, accessed 1 June 2022. https://blog.canberradeclaration.org.au/2020/05/08/climate-a -biblical-view-vs-the-worlds-view/

Smith E (2020) 'White Australia alone?: the international links of the Australian far right in the Cold War era', in Geary D, Sutton J and Schofield C (eds) *Global white nationalism*, Manchester University Press, Manchester.

Sovacool BK (2014) 'Cornucopia or curse? Reviewing the costs and benefits of shale gas hydraulic fracturing (fracking)', *Renewable and Sustainable Energy Reviews*, 37: 249–264.

Sparrow J (2021) 'Interviewing the far right is bad, so why do journalists keep doing it?: "no platform" from above and below', *Australian Journalism Review*, 43(2): 177–191.

Spash CL (2010) 'The brave new world of carbon trading', *New Political Economy*, 15(2): 169–195.

Spohr D (2017) 'Fake news and ideological polarization: filter bubbles and selective exposure on social media', *Business Information Review*, 34(3): 150–160.

Statista (2023) 'Countries with largest liquefied natural gas (LNG) export capacity in operation worldwide as of July 2022'. https://www.statista.com/statistics/1262074/global-lng-export-capacity-by-country/#:%7E:text=Australia%20and%20Qatar%20are%20currently,of%2071.6%20million%20metric%20tons.

Stephan K (7 January 2022) 'Three responses to climate change: there are basically three kinds of things we can do about global warming – Mitigation, adaptation, and suffering', *Mercatornet*, accessed 1 January 2023. https://mercatornet.com/climate-change-three-responses/76726/

Stone D (1991) 'Old guard versus new partisans: think tanks in transition', *Australian Journal of Political Science*, 26(2): 197–215.

Stutzer R, Rinscheid A, Oliveira TD, Loureiro PM, Kachi A and Duygan M (2021) 'Black coal, thin ice: the discursive legitimisation of Australian coal in the age of climate change', *Humanities and Social Sciences Communications*, 8(178), doi.org/10.1057/s41599-021-00827-5

Sydee J and Beder S (2006) 'The right way to go? Earth sanctuaries and market-based conservation', *Capitalism Nature Socialism*, 17(1): 83–98.

Taylor J (2020) 'The political economy of Australia's wildfires', *Harvard Political Review*, accessed 1 June 2022. https://harvardpolitics.com/the-political-economy-of-australias-wildfires/

Teaiwa K (2019) 'No distant future: climate change as an existential threat', *Australian Foreign Affairs*, 6: 51–70.

The Saltbush Club (n.d.) 'Saltbush solar activity watch established', accessed 1 January 2023. https://saltbushclub.com/2018/12/27/saltbush-solar-activity-watch-established/

The Spectator Australia (9 July 2022) 'All in it together', accessed 1 January 2023. https://www.spectator.com.au/2022/07/all-in-it-together/

Thomas T (20 October 2020) 'The climate cult's brat brigade', *Quadrant*, accessed 1 January 2023. https://quadrant.org.au/opinion/doomed-planet/2020/10/the-climate-cults-brat-brigade/

Timbo (29 July 2022) 'Australian Greens support same farming policies that crippled Sri Lanka and want us to starve', *Australian National Review*, accessed 1 January 2023. https://www.australiannationalreview.com/state-of-affairs/australian-greens-support-same-farming-policies-that-crippled-sri-lanka-and-want-us-to-starve/

Tobin P (2017) 'Leaders and laggards: climate policy ambition in developed states', *Global Environmental Politics*, 17(4): 28–47.

Traber M (1985) 'Alternative journalism, alternative media', *Communication Resource*, 7: 1–4.

University of Sydney (13 December 2021) 'Black Summer bushfire season cost farmers up to \$5 billion'. https://www.sydney.edu.au/news-opinion/news/2021/12/13/black -summer-2019-20-bushfires-cost-farmers-5-billion-australia.html.

van der Plank S, Walsh B and Behrens P (2016) 'The expected impacts of mining: stakeholder perceptions of a proposed mineral sands mine in rural Australia', *Resources Policy*, 48: 129–136.

Van Oldenborgh GJ, Krikken F, Lewis S, Leach NJ, Lehner F, Saunders KR, Van Weele M, Haustein K, Li S, Wallom D and Sparrow S (2021) 'Attribution of the Australian bushfire risk to anthropogenic climate change', *Natural Hazards and Earth System Sciences*, 21(3): 941–960.

Verschuer R, Melville-Rea H and Merzian R (2021) *What is Australia bringing to COP26?* The Australia Institute. https://australiainstitute.org.au/wp-content/uploads/2021/10/ COP26-Brief-The-Australia-Institute-WEB.pdf

Wahlquist C (13 September 2015) 'Peter Dutton apologises for "water lapping at your door" jibe', *The Guardian*, accessed 1 June 2022. https://www.theguardian.com/ australia-news/2015/sep/13/peter-dutton-apologises-for-water-lapping-at-your-door -jibe

Wardle C and Derakhshan H (2017) *Information disorder: toward an interdisciplinary framework for research and policymaking*, Council of Europe, Strasbourg.

Walters R and Martin P (2016) 'Trade in "dirty air": carbon crime and the politics of pollution', in R White, M Kluin and T Spapens (eds) *Environmental crime and its victims: perspectives within green criminology*, Routledge, London.

Walther S and McCoy A (2021) 'US extremism on Telegram', *Perspectives on Terrorism*, 15(2): 100–124.

Weber D, Nasim M, Falzon L and Mitchell L (2020) '#ArsonEmergency and Australia's "Black Summer": polarisation and misinformation on social media', *Proceedings of Multidisciplinary International Symposium on Disinformation in Open Online Media*, pp. 159–173. Springer, Cham.

Wells S (2023) 'Stephen Wells', *XYZ*, accessed 1 January 2023. https://web.archive.org/web /20220729015952/https:/xyz.net.au/author/stephen-wells/

Westmore P (31 August 2022) 'Renewable energy is dirty and unethical', National Civic Council, accessed 1 January 2023. https://ncc.org.au/newsweekly/energy-science-enviro /renewable-energy-is-dirty-and-unethical/

Williams G (13 June 2019) 'A remedial lesson in climate education', *Quadrant*, accessed 1 January 2023. https://quadrant.org.au/opinion/doomed-planet/2019/06/a-remedial -lesson-in-climate-education/

Wilson R (28 June 2022) 'Globalists want to replace children with virtual computer fakes', *Australian National Review*, accessed 1 January 2023. https://www.australiannatio nalreview.com/lifestyle/globalists-want-to-replace-children-with-virtual-computer -fakes/

Wood MA, Richards I and Iliadis M (2022) *Criminologists in the media: a study of newsmaking*, Routledge, London.

World Meteorological Organization [WMO] (2021) *Climate change indicators and impacts worsened in 2020*. https://public.wmo.int/en/media/press-release/climate-change-indicators -and-impacts-worsened-2020

Yemini A (2022) 'Climate alarmists must think we have a very short memory' [YouTube], accessed 1 January 2023. https://www.youtube.com/watch?v=95JsmqFNDNw&ab _channel=AviYemini

Zappone C (12 February 2020) '#ArsonEmergency: how "fake news" created an information crisis about the bushfires', *Sydney Morning Herald*, accessed 1 June 2022. https://www.smh.com.au/world/oceania/arsonemergency-how-fake-news-created-an-information-crisis-about-the-bushfires-20200211-p53zma.html.

Zheng Z and Bhatt B (2022) 'Political polarization in Australia: a case study of brushfires in Australia', in Qureshi I, Bhatt B, Gupta S and Tiwari AA (eds) *Causes and symptoms of socio-cultural polarization*, Springer, Singapore.

Zhou N (14 November 2019) 'Former Australian fire chiefs say Coalition ignored their advice because of climate change politics', *The Guardian*. https://www.theguardian.com/australia-news/2019/nov/14/former-australian-fire-chiefs-say-coalition-doesnt-like-talking-about-climate-change

5 New Catastrophism and the Environment–Security–Development nexus

Programming and advocacy during the climate crisis

Introduction

Climate change modelling reveals the devastating ecological and environmental impacts of more than 1.5°C of global surface temperature warming from pre-industrial levels. Along with sea-level rises and the collapse of local ecosystems, the many societal impacts of climate change include 7 million 'environmental migrants' displaced by natural disasters in 2020, with many more projected to be displaced by 2050 (Pörtner et al. 2022). Far-right actors have also emerged in this context as a growing social crisis, capitalising on global popular concerns about the environment, while advocating authoritarian solutions to the climate crisis (Forchtner 2019). As explored in earlier chapters of this book, often referred to as 'ecofascist' are extreme population control measures advocated by right-wing activists and ethnonationalist governments, and the accelerationist propaganda of violent actors seeking to hasten the social and economic collapse of societies worldwide (Ross and Bevensee 2020; Moore and Roberts 2022).

This book has sought to unpack responses to climate change on the part of the extra-institutional far right in Australia, ranging from the blood and soil tropes of self-described ecofascists to historical Third Positionists, NAs and the proponents of various historical strains of Australian nativism. The book's analysis of news and social media has examined how the contemporary far-right social turn occurring both globally and within Australia not only relies on its proponents' quest for a radical instauration and protection of 'white power' but also feeds on wider societal recognition of the need for a radical transformation of human societies to stave off the many feared eschatological effects of global heating (Malm 2021).

While not all populationist or sustainability arguments may be understood as either 'ecofascist' or far-right, engagement with these arguments by the multi-spectrum right effectively constitutes a periphery of eco- or *ur-fascist* (Eco 1995) political organisation amenable to co-optation and entryism by a more groupuscular accelerationist and authoritarian ecofascist core (see Griffin 2018; Chapter 6, p. 216). To this end, we argue, political-institutional media centralising disparate discourses about climate change and the environment also represent an avenue for the greater mainstreaming of far-right environmentalist ideas (Richards et al. 2022).

DOI: 10.4324/9781003325437-5

This chapter accordingly emphasises how the far right's use of the environment occurs within an ideological environment characterised by increasing recognition among both radical social movements and more mainstream political institutions of likely future climatic impacts. Warnings that climate change may lead to ecological and social catastrophe have grown more urgent from mainstream environmental institutions, such as the IPCC, the WMO and national academies of science in many countries (Monbiot 2007; Hansen 2009). The response to such warnings has been an outpouring of doomsday literature and proposals for systemic change, representing a new 'epoch' in environmental thought sometimes labelled 'New Catastrophism' (Urry 2011, 36).

The analysis in this chapter does not dismiss as either erroneous or excessive the warnings of New Catastrophists or others calling for radical and transformative responses to climate change. Rather, it presents a critical, but hopefully constructive reflection on radical perspectives on transformative change in the way human societies are organised that provide alternatives to the far right. This investigation attends to the role of global capitalism in shaping the dialectic relationships between different climate responses, and the discussion emphasises the risks and tensions that can arise through far-right co-optation and entryism into different climate change response scenarios. In this chapter, we ultimately seek to develop further upon the radical responses to climate change posited by New Catastrophists and others, to sketch out possibilities for mitigatory climate action that represents tangible and increasingly necessary alternatives to the ethnonationalism and authoritarianism of the far right.

For the purposes of this book's investigation, we note that Australian far-right actors' heterogeneous uses of climate change must be contextualised in relation to the wider political, oppositional field within which they operate, and through reference to which these actors draw propagandising support. To this end, the first part of the discussion examines what has been termed the ESD nexus in intergovernmental responses to the security threats related to climate change, political violence and matters of economic–human development. In this context, the investigation considers the propagative effect of securitising ESD discourses about the effects of global heating on migration, resource scarcity and intra-state conflict for both global heating and far-right political advancement.

As the combined analysis seeks to show, both reactionary and progressivist climate responses operate within a wider remit of recognition about the global impacts of presently unfolding forms of ecological-environmental devastation. Building on the previous insights and findings set out in this book, this chapter demonstrates how far-right environmentalist expressions are far from an isolated phenomenon. As discussed in the introductory chapter to this book, they reflect the dominant global political–ideological environment in which media and political actors in the Global North have erroneously placed blame for climate change on the Global South through rhetoric about population control and fossil fuel usage (see Thomas and Gosink 2021; Moore and Roberts 2022).

The discussion therefore begins by examining international institutional responses to the intersections between political violence and climate change in relation to the ESD nexus. This discussion contextualises the rise of far-right environmentalist politics within global developmentalist historical conditions that securitise Global South populations detrimentally affected by global heating, and it considers the ways in which institutional responses to climate change and the environmentalist nexus can both pattern and inform the contemporary far right. Building upon a growing awareness of the need for transformative structural responses to climate change within both institutional and intergovernmental arenas, the second section discusses the tendencies of New Catastrophism (Urry 2011, 36), outlining a framework for understanding the different currents of radical response to climate change that variously resist and in some ways provide an ideological backdrop to contemporary far-right environmentalist expressions. The analysis concludes by addressing New Catastrophism's rootedness in disparate historical traditions of response to environmental issues by progressivist eco-socialist and eco-anarchist political tendencies.

The investigation in this chapter is pursued not for the purpose of drawing moral equivalence between disparate 'right' and 'left' radical traditions. Rather, this analysis seeks to excavate the social and political forces that can differentially reinforce, serve as an ideological backdrop for, or provide alternatives to the multi-spectrum far-right environmentalist positions examined in the preceding chapters.

The Environment–Security–Development nexus

The ESD nexus has predominantly been examined through a somewhat insular lens, which often overlooks the connections between various forms of non-state political violence in both Global North and Global South settings, as well as the broader conditions that shape global trends in political violence (Bloom 2016; Jackson 2018). Likewise, the harmful and crime-generating actions of violent state actors in militaristic conflicts have been inadequately acknowledged and addressed (Peoples and Vaughan-Williams 2020).

Discussions of non-state political violence associated with climate change usually focus on the Global South and LDCs. However, far-right actors in the Global North are increasingly exploiting the perceived consequences of climate change, referencing purported resource scarcity, migration and environmental degradation (Forchtner 2019). Concurrently, neo-jihadist groups like Al-Qaeda and Islamic State frequently capitalise on neo-colonial development histories that negatively impact LDCs (Richards 2020b). These global trends in political violence transpire against a backdrop of environmental security policies and initiatives that highlight the national and international security risks posed by climate change stressors to affluent nations (Richards 2023).

The securitising approach to the ESD nexus has effectively raised the issue of environmental security to the level of high politics, where it can ostensibly receive greater resourcing and attributions of significance (Warner and Boas 2019). This shift has also, though, yielded the undesirable effect of assigning primary responsibility for dealing with the prominent climate impacts on people and territories to the hard mechanisms of intra-state security, supported at intergovernmental arenas (Busby 2018). This has occurred despite the fact that LDC populations experience the gravest impacts of environmental change, while, as explored in several parts of this book, countries primarily in the Global North have historically been the greatest per-capita contributors to climate change (Hickel et al. 2022).

The potential drawbacks of more stringent security strategies in addressing climate change and its security implications stem from the elevation of climate security beyond risk assessment to a realm that seemingly justifies urgent anti-democratic actions aimed at controlling vulnerable populations (Thomas and Warner 2019). These actions may encompass the suspension of norms, regulations and other legal frameworks within Schmidtian 'states of exception' (Mehta et al. 2019). However, these tendencies should not be viewed as exclusive to non-liberal or overtly authoritarian states. In fact, historically, and more recently during the COVID-19 pandemic, liberal democratic states have demonstrated a capacity and willingness to adopt illiberal and authoritarian measures in response to crises (Brinn 2022). Despite stated concerns for maintaining stability and preserving order, a multi-level techno-authoritarian politics that focuses on the need to manage resource scarcity and global population movement due to climate change could itself become a potential catalyst for interstate and regional conflicts. The apprehensions that underlie authoritarian approaches to climate and security governance are often based on predictions made since the 1990s of future disastrous interstate conflicts driven by resource scarcity, such as the 'water wars' debate (McDonald 2018; Mehta et al. 2019).

The ESD literature has predominantly focused on regional insurgencies and non-state violence in sub-Saharan Africa and the Middle East, which are said to be linked to climate change. Frequently mentioned examples include al-Shabaab acting as an alternative aid and service provider in Somalia following droughts and floods; Boko Haram recruiting vulnerable individuals affected by drought, deforestation and food scarcity in the Lake Chad region of Central Africa (Szenes 2021); and Islamic State leveraging water and wheat as weapons of war to recruit and control populations in Iraq and Northern Syria.

This focus on non-state actors exploiting climate change for specific political objectives is sometimes justified by statistics on terrorism incidents, such as the fact that five countries account for 72% of terrorism-related deaths (Iraq, Afghanistan, Nigeria, Pakistan and Syria), while four groups are responsible for 74% of deaths (Islamic State, Boko Haram, Al-Qaeda and the Taliban) (IPI 2022). However, this narrow emphasis can sometimes overshadow the regional interests and power dynamics at play in conflict zones across Central Asia, Africa and the Middle East. Among numerous historical examples and cases, these interests were apparent in Russian and Iranian support for the Bashar al-Assad government from 2011, as well as Qatar and Saudi Arabia's backing of jihadists in the ensuing civil war (Linke and Ruether 2021). Foreign powers' involvement was also evident in US support for both Kurdish forces and Al-Qaeda (the former Jabhat al-Nusra branch) in Syria during this period, where Al-Qaeda's weapons were later seized by Islamic State (Richards 2020a).

Programming for localised violence prevention initiatives has been supported by the UN in Central Asia, Iraq and Sudan. A recent policy framework and rationale guiding these interventions is set out in a 2020 United Nations Development Programme (UNDP) brief titled 'The climate security nexus and the prevention of violent extremism: working at the intersection of major development challenges'. Citing 'lessons learned from a climate security perspective of efforts to prevent VE [violent extremism] in politically and environmentally fragile contexts affected by climate change', the brief emphasises the importance of 'holistically considering the risk landscape' through attention to 'convergence between climate and security risks' in order to avoid 'maladaptation' (UNDP 2020). A wider ESD programming literature applies a similar intersectional focus, emphasising the need to strengthen adaptive migration measures in response to climate impacts through education, the facilitation of property rights, and by supporting knowledge flows and technology transfers between regions (Mach et al. 2019). Macro-structural attention to the role of development and economic systems in generating global heating is typically under-addressed (McDonald 2018).

By contrast with efforts to address climate-related conflict in the Global South, there has been relatively scant intergovernmental attention paid to the exploitation of the effects of climate change by far- and extreme-right actors, though community-based initiatives are more common. One such community-based initiative was the Kungälv model of education in western Sweden, which sought to counter 'harassment, bullying and violence' at a Ytterby school, which the environmentalist neo-Nazi Nordic Resistance Movement had previously used as a recruiting ground (Szenes 2021). Another example was the Resilient Vermont Network, which broadly sought to build local community resilience to the risks posed by natural hazards (Szenes 2021). With funding from the Association for the Study of Literature and Environmental Leadership, The Institute for Ethics and Public Affairs, and San Diego University, another effort in 2022 included members of the Anti-Creep Climate Initiative developing educational materials for delivery in schools and universities, in the form of a zine featuring Marvel superhero characters titled 'Against the eco-fascist creep: debunking ecofascist myths' (Anson 2022).

The relatively atomised intergovernmental and civil society ESD focus on climate-related violence at a substate level and in developing regions reflects the alignment of resourcing with the UN's historical protection of state interests and priorities, as set out in 2007 and 2011 United Nations Security Council (UNSC) debates on the merits of securitising climate change. In a 2007 UNSC debate on whether the council should play a role in global responses to climate change, for example, several Global North state representatives agitated for global security responses to threats to stability caused by migration, while speakers from less developed regions such as Namibia and Papua New Guinea (speaking on behalf of the Pacific Islands Forum) likened climate change to a form of 'warfare', citing the climate impacts for them as akin to the types of 'bombs' and 'guns' feared by developed states (Detraz and Betsill 2009). While Namibia and Tuvalu compared 'greenhouse gases to chemical warfare', Tuvalu described 'chimney stacks and exhaust pipes as weapons' (Detraz and Betsill 2009, 311).

The existing ESD focus also reflects the limitations of an intergovernmental security and development apparatus for responding holistically to climate change's destabilising effects, beyond awareness of its nation-state mandate (McDonald 2013; 2018). The conflicting aims inherent in the concomitant approach to 'development', 'security' and 'environment' under the ESD nexus, as evident in many cases, can be seen to relate to their differential aims. Coarsely put, development seeks (economic and societal) growth; security seeks stability; and environment, as emphasised in an emergent ecological security literature, requires transformative change in the reorganisation of human societies to meet the scale of risk posed to human and non-human life from unfolding forms of ecological and environmental change (see McDonald 2013; 2018).

Notwithstanding these tensions, contemporary political debate on climate change has led to concepts such as the Anthropocene planetary boundaries and to the ESD nexus featuring extensively in discussions about the relationships between interstate or non-state political violence and environmental security (Mehta et al. 2019). In contexts where national and international security concerns predominate, these are sometimes supplemented by a 'human security' focus on individual well-being and the capacity of national populations for climate resilience. Human security responses to the ESD at the United Nations General Assembly (UNGA), the United Nations Environment Programme (UNEP) and UNDP settings for intergovernmental policy-making, conceptualised as 'secure human development' (Murphy 2015), tend to foreground the individualised interests and abilities of states and multilateral institutions as a key vehicle for adaptive and mitigatory climate action (Busby 2018). These tendencies, however, are operating within the context of a rapidly evolving broader public conversation about climate responses. As we will see in the next section, the more public-facing New Catastrophist literature draws on similar themes and science but advocates for radical social and political change rather than adaptation of the status quo, directing its recommendations towards general populations, who are often portrayed as the primary agent of change (e.g. Hansen 2009; Foster 2013; Klein 2014).

The focus on human security in ESD debates conversely relates to strategic state priorities set out in the UN's 2006 Global Counter-Terrorism Strategy. As the first of its kind, this strategy reflects Amartya Sen's capability-based approach to development enshrined in the 2000 Millennium Development Goals (MDGs), insofar as it is committed to social progress and human rights in place of the post-9/11 US-directed hard-security context of the global war on terror (Richards 2020b). The intersections between the counter-terrorist foreign interventions of powerful states and the long-term sequelae of development actions over time, however, have exacerbated the structural economic and environmental factors contributing to contemporary forms of non-state violence (Weis and White 2020). The long-term effects of development programmes led by multilateral institutions in which Global North states yielded the greater influence within LDCs have also been criticised for the lack of institutional accountability they were seen to promote, in particular regarding the medium- and long-term impacts of development projects on the environment (Floyd 2015). Moreover, for New Catastrophists and others, this focus on state capacities and economic development is increasingly seen to bracket the possibilities for more fundamental societal transformations to address global heating as a root cause of violence, instability and social harm (see e.g. Foster 2009; Parenti 2011; McDonald 2018).

The risks of maladaptation in securitising climate change and the encroachment of security positions on developmental responses to matters of human safety have then in some cases led to the disproportionate attribution of risk to the civil instability wrought by climate change activism, in light of the extent of the collective suffering of international communities as a result of extreme environmental devastation. In addition to the securitisation of populations in developing states, this phenomenon has also manifested within the Global North. One notable example is the controversial UK counter-terrorism police listing of the nonviolent environmental activist group Extinction Rebellion as an extremist organisation. This designation has meant that the group's members could be referred to authorities responsible for managing the contentious terrorism monitoring programme, Prevent (Berglund and Schmidt 2020). The inclusion of a nonviolent environmental group in such a list highlights the extent to which securitisation efforts have expanded, raising concerns of a backlash against progressivist environmental activism, and the potential for a growing authoritarianism, which Malm (2021) refers to as 'fossil fascism'.

Notwithstanding the limitations of ESD approaches, the 2022 IPCC 'Mitigation of climate change' report was seen by some as marking a turning point away from previous UN communications on climate change by virtue of its instructive recommendations on possible transformative long-term mitigation scenarios. The summary paper outlined means and methods by which the well-being measures in the 2030 SDG agenda might be meaningfully achieved by promoting elements of degrowth, no growth or other post-capitalist economies in developed countries, or a signposting of other scenarios in which sustainability incentives may be replaced with other measures of social and economic progress. The conceptual degrowth ideas highlighted in the report, for example, include a focus on forms of prosperity not reliant upon growth and the societal promotion of non-material values (Parrique 2022). The technical degrowth notions discussed include a decline in income levels in wealthy states (or a maximum wage); minimum and maximum standards of consumption, such that it does not impair prosperity or the 'good life'; the introduction of universal basic incomes; incentives for eco-innovation; and community-managed energy (Parrique 2022; Pörtner et al. 2022).

More broadly indicating this appetite for change, a radical rethinking of methods of societal governance has also in some cases been platformed within governmental settings. For example, an article by Spain's Minister of Consumer Affairs and leader of the United Left party, Alberto Garzón, titled 'The limits to growth: eco-socialism or barbarism', warned of a 'reactionary, ecofascist solution to the eco-social crisis', and a need to downscale economic activity to levels conducive to natural planetary boundaries (Garzón 2022). In the wake of the 2022 IPCC report,

UN Secretary-General António Guterres posted on Twitter: 'climate activists are sometimes depicted as dangerous radicals. But the truly dangerous radicals are the countries that are increasing the production of fossil fuels. Investing in new fossil fuels infrastructure is moral and economic madness' (@antonioguterres 2022).

As outlined in the following section, the extensive and expanding New Catastrophist literature relies heavily on mainstream environmental and climate science institutions, particularly the IPCC, to substantiate its grim predictions and radical proposals. Based on increasingly alarming reports from the first decade of this century, New Catastrophists advocate for swift and systemic social, political and economic transformation – a necessity that they argue, and we concur, is implied in the longstanding warnings of the IPCC and others (Hillman et al. 2007, 21; Shearman and Smith 2007, 5–6; Spratt and Sutton 2008, 45; Campbell 2009, 49; Leahy et al. 2010, 853; Gilding 2011, 33; Parenti 2011, 5). These latest statements align such institutions explicitly with the New Catastrophist literature, demonstrating that the transformations they demand are neither alarmist nor fanciful but rather a logical extension of climate change reality into the social and political spheres. While the early New Catastrophist literature made explicit what was arguably implicit within climate science, the climate science community has since openly reached similar conclusions, rendering the ESD focus on state capacity and economic development inadequate (IPCC 2021; 2022; 2023).

Statements emphasising the gravity of the impacts of global heating and its industrial and economic causes have been met by lofty agency benchmarks of reparations for global populations disproportionately suffering from unequal development and persistent inequality exacerbated by climate change. Such aims were outlined in a statement of values for violence prevention and climate change mitigation within a 2021 UN report released by Secretary-General Guterres, 'Our common agenda'. The report included a section titled the 'New agenda of peace', which set out five policy prescriptions related to addressing inequality: providing at least USD100 billion per year to developing countries for climate action; improving analysis of warning systems; developing partnerships and initiatives linking local, regional and national approaches; and ensuring sustained investment in climate change mitigation and response (UN 2021). The Zurich Flood Resilience Alliance found, however, that the majority of countries most vulnerable to climate change received less than USD20 per person yearly for climate change adaptation over the period 2010–2018, while nationally determined contributions detail that most developed countries' GHG reduction targets remain insufficient to meet the Paris Agreement target of limiting temperature increases to below 2°C (Wong et al. 2020).

Some researchers and practitioners from within academia, civil society and government might err towards transformative or structural change intended to mitigate the security and conflict risks that climate change generates, but it is only recently that this approach has been represented beyond the margins of environmental security discussions. As explored in greater detail in the following section, growth-critical and other more transformative theorisation regarding climate change mitigation and response has more often emerged from radical environmentalist positions than from within institutions; indeed, the number of peer-reviewed articles about 'degrowth', including articles with some variation in terminology, numbered only three in 2007, relative to over 340 in 2022 (Parrique 2022).

It is also important to emphasise that, despite the greater engagement between institutional actors and activist communities, the climate impacts on disenfranchised populations remain primarily conceptualised by policy communities as 'risk' or 'threat' multipliers, whereby the shocks and stressors of acute and chronic weather events displacing and disrupting people in the Global South are frequently portrayed as an object for securitisation, particularly by actors from the Global North (Abrahams 2019). This 'ontological politics' of interpreting climate-affected social relations as 'agonistic' can be used to justify states' preparations for realist or 'realpolitik' conditions of endless cycles of violence, which are at once both performative and self-fulling (see Barnett 2019). Forced migration is also often a key issue in point, both within and between states and regions, and from LDCs towards countries that developed historically via extractivist methods of global production, and which now institute often highly restrictive (if not also symbolically and materially violent) border protections (Richards 2020b).

The objectification of Global South populations is not confined, however, to the more technocratic ESD discourse, but also pervades the ostensibly more radical and humanitarian New Catastrophist thought. In New Catastrophist narratives, the increased threat of climate disruption and displacement is presented less as a reason for securitisation and more as a catalyst for engaging in the actions necessary for effective climate action achieving radical social change and broader environmental sustainability. The focus on extreme social consequences in the Global South can be interpreted as an attempt to accurately depict the threat faced and those most likely to experience it. However, a sceptical reader might suggest that since one of the primary goals in most New Catastrophist literature is to motivate its target audience – the general populations of the affluent Global North – to undertake the specific actions prescribed by New Catastrophism as essential for survival, such narratives could also function as an implicit warning that plays on fears of immigrant 'inundation' and non-white 'replacement' that ecofascists emphasise (Dyer 2010, 18–21; Foster et al. 2011, 439–440; Parenti 2011, 8–9, 237; Steffen and Griggs 2013, 134–138).

In the context of debates on the prevention of extra-state violence, there is also a paradox inherent in some climate change mitigation measures implemented by international powers within the Global South, given the anti-immigration and ethnonationalist political expressions in response to climate change seen in wealthier states. For example, the Hungarian far-right organisation Our Homeland Movement refers to itself as a 'green party' with a 'green wing', claiming to protect Hungary from 'foreign land grabs' (Szenes 2021). Simultaneously, although beyond the scope of this book, the occasionally detrimental impacts of wartime development activity by Northern actors in LDCs and contemporary peacetime greening efforts in the Global South illustrate how climate change mitigation measures could similarly be another chapter in the extended history of neo-colonial interventions (Mousseau 2019). As the preceding discussion has attempted to demonstrate, not only are intergovernmental and community-led actions against the environmentalist far right limited, but there has also been a lack of attention paid to the intersecting historical factors arising from developmentalism and economic growth that affect both global heating and the rise of the far right.

New Catastrophism

Acknowledgement of the catalysts for climate change and its social impacts also accompanies recognition of the need for transformative change to address its physical and social risks. This recognition has emerged within New Catastrophist circles partly in response to scientific revelations emphasised recently by the IPCC and other intergovernmental institutions. The latest 2023 IPCC report warned that the world will reach over 1.5° of warming above pre-industrial levels within the next two decades unless drastic emission cuts are implemented, with a 4° increase by the end of the century being likely if states continue to pursue high-carbon pathways. Human activity causing GHG emissions is projected to lead to melting glaciers, intense heatwaves and warming oceans. Uniquely, compared to earlier IPCC and UNEP statements, the documents released since 2021 also emphasise that at current emission levels, the world is already close to various 'tipping points' at which irreversible damage could lead to global extinction-level events, including perpetual sea-level rises and scenarios in which forests would cease growing (IPCC 2021; 2022; 2023).

The 'new' in New Catastrophism emphasises this contemporary awareness of climate change effects, distinguishing the current wave of global heating responses from the 'old' catastrophism associated with alarmist environmentalist texts. This earlier wave included scarcity debates responding to the Club of Rome's 1972 'limits to growth' study (see Meadows et al. 1972), and caution about the supposed deleterious effects of 'overpopulation' in the Global South popularised through texts such as Paul Ehrlich's *The Population Bomb* (1968) (see Chapter 1, p. 4). Old catastrophism also translated into radical climate solutions that were vulnerable to exploitation by far- and reactionary-right social groups, such as William Ophuls's authoritarian environmentalist proposals in his *Ecology and the Politics of Scarcity* (1977). Although overpopulation remains a primary issue for some environmentalists, New Catastrophists tend to highlight the anthropogenic factors contributing to climate change beyond the population size. They draw attention to the role played by over-consumption, the burning of fossil fuels, the mining of essential minerals and commercial activities incommensurate with a sustainable planet, such as animal agricultural practices and deforestation destructive to the natural environment.

Some New Catastrophist theories, however, also provide an ideological narrative that is vulnerable to exploitation by climate change accelerationists or a resignatory and ethnonationalist far right. More broadly, some New Catastrophist texts might be said to foster an ideological milieu within which far- and extreme-right social forces exploiting climate change may thrive. We do not suggest by any means that this is the active or passive intention of New Catastrophists. Moreover, much of the New Catastrophist literature is non-academic and is therefore often more polemical, journalistic or activist in style, designed to raise awareness of devastating climate impacts, and so it should be viewed on its own terms. We believe that our critique, though, is justified by the fact that New Catastrophist perspectives make a significant high-profile contribution to global public political debate on global heating-related issues.

In the interests of understanding the public impact of New Catastrophist narratives, we suggest it is worth considering the potential far-right co-optation of New Catastrophist narratives that intersect with 'collapse' theories; particularly those alleging that global heating will, in the near future, lead to devastating effects for humanity, including the potential breakdown of civilisation and even the extinction of the human race (Urry 2011). Notwithstanding its eschatological focus, the essential characteristic of the New Catastrophist literature is the warning of a looming catastrophe, with the hope of motivating preventative action (Lilley et al. 2012). Unlike the scientific texts, which only describe the situation (e.g. Kolbert 2014), New Catastrophists 'are all concerned with … the systemic restructuring of economic and social life' (Urry 2011, 46) (e.g. Foster 2009; Klein 2014, 450–457). The solutions proposed by New Catastrophists mostly also derive from radical left environmentalist positions, contrasting with right-wing protectionist or xenophobic climate responses. These proposals for social transformation can thus range from anarchistic ecotopian social visions (Trainer 2011) to authoritarian state regimes (Shearman and Smith 2007), rendering some approaches more amenable to co-optation by the authoritarian far right than others.

New Catastrophist political interventions have the capacity to appeal to extra-institutional social movements partly by virtue of their extra-academic orientation. They represent first and foremost an attempt to inspire political action and bring about social and political change. Many New Catastrophist texts are thus aimed at a popular audience, and their chief proponents variously have backgrounds in environmental and social activism. New Catastrophism has therefore been driven by such ideological dispositions as 'dire realism' (Foster 2013, 9); 'climate realism', representing the position that 'there is no escaping the science' of climate change (Spratt and Sutton 2008, xii; Orr 2009, xvi, 196; Hamilton 2010, 1); and 'ecological realism' (Gorz 1980, 11), the idea that infinite economic growth is incommensurable with continued human survival (Shearman and Smith 2007, 14; Orr 2009, 196; Hamilton 2010, 53; Klein 2014, 21).

According to New Catastrophist thought, we live on a *Suicidal planet* (Hillman et al. 2007), an *Apocalyptic planet* (Childs 2013), where *Compounding crises* (Steffen and Griggs 2013) look set to deliver *A rough ride to the future* (Lovelock 2014). *Any way you slice it* (Cox 2013), it is *Climate code: red* (Spratt and Sutton 2008). We are on a *Countdown* (Weisman 2013) to a *Carbon crunch* (Helm 2012) and *Climate cataclysm* (Campbell 2009) that threatens a future of *Heat* (Monbiot 2007), *High tides* (Lynas 2005), *Meltdown* (Gow 2009), *The erosion of civilizations* (Montgomery 2012) and *An uninhabitable planet* (Wallace-Wells 2020). *When the rivers run dry* (Pearce 2006), and other resources such as food and energy become scarce, we will endure *Climate wars* (Dyer 2010) and possibly *The next world war* (Woodbridge 2004). Nonetheless, we are blithely going *Down to the wire* (Orr 2009), *Ignoring the apocalypse* (Davis 2007) and *Stumbling towards collapse* (Leahy et al. 2010).

We must recognise that *The party's over* (Heinberg 2005) and that climate change *Changes everything* (Klein 2014). To *Cancel the apocalypse* (Simms 2013) we must bring about *The ecological revolution* (Foster 2009) by *Making peace with the earth* (Shiva 2013). It will not be easy – we will need to endure a *Great disruption* (Gilding 2011) and *Long emergency* (Kunstler 2007) to succeed in *The big pivot* (Winston 2014) to an *Alternative world system* (Baer 2012) based on a new *Deep economy* (McKibben 2007) with a *No-growth imperative* (Zovanyi 2013). However, if we do not have the courage to undertake this difficult task, then there will be *Apocalypse forever* (Swyngedouw 2010). We will see *The end of ice* (Jamail 2019) and *The collapse of western civilization* (Oreskes and Conway 2013), and we will enter *The last hours of humanity* (Hartmann 2013) before finally ourselves becoming victims of *The sixth extinction* (Kolbert 2014).

Notwithstanding the apocalyptic narratives on display in many New Catastrophist texts, solutions are often presented as a realistic response to the mainstream scientific consensus regarding climate change. Mainstream plans for global heating mitigation and carbon emission reduction targets have long been informally based on estimates that 2°C above Pre-Industrial Averages (PIA) represents a 'safe level' of warming, below which tipping points were thought unlikely to occur (Nordhaus 1977; Spratt and Sutton 2008, 89–98). This commitment was formalised in the 2015 Paris Agreement with the aim expanded to 'well below' 2°C 'and pursuing efforts to limit the temperature increase to 1.5°C above' PIA (IPCC 2018). However, the latest science suggests that both the 'guardrail' of 2°C, and the more ambitious 1.5°C limit, which, on our current trajectory, we are expected to exceed, are still too high, as some feedback processes have already begun at our current warming level of 1.15°C above PIA, at the time of writing (WMO 2022). Therefore, many New Catastrophists suggest that a more realistic target to avoid dangerous tipping points and runaway change is, at most, 1°C above PIA (Hansen 2009, 74).

Although New Catastrophists refer to the IPCC to establish the reality of scientific consensus on the basic outline of the threat, they frequently argue that the IPCC is too conservative, and that the problem is worse than even the dire scenarios it outlines (Spratt and Sutton 2008, 65–70; Hansen 2009, 74–76; Hamilton 2010, 3). One reason for this is the IPCC's reluctance to include in its modelling the effects of 'system feedbacks' and 'tipping points', which are central concerns for New Catastrophism (e.g. Foster 2009, 44; Hamilton 2010, 25; Urry 2011, 40).

Exemplary here is a report released by the UN in May 2022, although underreported in international broadcast media, titled 'Global assessment report on disaster risk reduction' (the GAR – UNDRR 2022) and endorsed by UN Secretary-General Guterres, which drew on a paper published by the UN Office for Disaster Risk Reduction (UNDRR) and authored by Thomas Cernev of the University of Cambridge's Centre for the Study of Existential Risk. The GAR stated that at least four of nine planetary boundaries currently appear to be operating outside the safe limits for human habitation. Moreover, an independent analysis of these findings identified that, contrary to the GAR's official findings, at least six boundaries seem to have already been surpassed (Ahmed 2022). The UNDRR report upon which much of the GAR was based also outlined that in three of four likely scenarios for climate change mitigation, and short of drastic action to meet emissions reduction targets, 'Global Catastrophic Risk Events' – defined as leading to either 10 million fatalities or greater than USD10 trillion in damages – combined with the breaching of planetary boundaries would lead to a reduction in safe operating spaces for human societies such that 'total societal collapse is a possibility' (UNDRR 2022; Ahmed 2022).

Perhaps the most cited source in the New Catastrophist literature in this respect, aside from the IPCC, is James Hansen, a highly respected climate scientist whose work on tipping points (e.g. Hansen 2007; Hansen 2009, 149) is widely drawn upon (Spratt and Sutton 2008, 125; Foster 2009, 58; Orr 2009, 184; Hamilton 2010, 13; Gilding 2011, 128; Lynas 2011, 57; Parenti 2011, 60). The science of tipping points is very uncertain; they are left out of the IPCC's predictive climate models, given that 'models differ considerably in their estimates of the strength of different feedbacks in the climate system' (IPCC 2007, 67; Leahy et al. 2010, 853). Hansen's warnings, however, are not based on predictive models but on historical comparison with geological records of periods with CO_2 concentrations similar to the current levels.

Based on geological records, Hansen asserts that the current concentrations of CO_2 and other warming gases in our atmosphere are approaching, and are predicted to reach by 2100, the level associated with a temperature rise of at least 4°C (Hansen 2009, 117, 160; Hamilton 2010, 21). This level is thought to be well beyond climate 'tipping points' that would trigger sudden dramatic events, creating self-reinforcing 'positive feedbacks' that accelerate the entire process. This state of affairs could then lead to further temperature rises of between 6°C and 10°C (Spratt and Sutton 2008, 49) and possibly as high as 11.5°C above PIA by 2100 (Monbiot 2007, 6, 13), which could trigger a mass extinction event that includes humanity (Hansen 2009, 236, 149). Previous similar occurrences, such as the Paleocene-Eocene Thermal Maximum (PETM) extinction event of approximately 65 million years ago and the end-Permian extinction event of approximately 251 million years ago, saw ice-free polar conditions, global sea levels approximately 70 metres above current levels and cascading effects that eventually wiped out an estimated 95% of all life on the planet (Hansen 2009, 146–150; Kolbert 2014, 103). Recent work by Hansen affirms his earlier estimates, arguing that even they were based on overly optimistic assumptions regarding the effect of already released climate gases, and that even if emissions were to rapidly cease now, current GHG levels look set to deliver a 10°C temperature rise above PIA at up to 20 times the rate of change before reaching equilibrium as that which characterised the PETM (Hansen et al. 2022).

Ecofascists on the extreme right tend not to draw explicitly on New Catastrophism agitating in response to climate science. As mentioned, however, we argue that it is important to recognise the role that discourse relating to such collapse scenarios plays in generating an ideological milieu within which extreme-right responses to climate change operate. Collapse scenarios regarding a 'burning world' in particular provide context for accelerationist eco- and neo-fascist advocacy of palingenesis, representing the eugenicist rebirth of a state along racial–national lines, from the ashes of the society that came before it (Thomas and Gosink 2021; Moore and Roberts 2022). We are not suggesting that the science should be obscured, but we do argue that it is prudent to recognise these potential effects.

There are also important resonances between the superficial right-wing narratives of Western civilisational decline and the more substantive New Catastrophist emphasis on the potential for climate tipping points to trigger dramatic social change that could lead to the breakdown of *human* civilisation. Drawing on the work of Jared Diamond (2005) and Joseph Tainter (1990) that examines the failure of such historical civilisations as that on Easter Island and the Maya civilisation, some New Catastrophist theory has been particularly motivated by a fear of 'collapse' (e.g. Foster 2009, 120; Gilding 2011, 107–108; Steffen and Griggs 2013, 122). Echoing realpolitik security concerns dominant in the ESD literature, the notion of *societal* collapse affecting human civilisations is seen by New Catastrophists as a process characterised by 'wars and civil conflict, famine, population crash', and so on (Leahy et al. 2010, 852). In a *climate* collapse scenario, New Catastrophists believe that humanity would be in danger of extinction due to our inability to prevent further 'runaway' climate change, and that this would lead to such consequences as the mass failure of food sources, widespread water shortages and changes in weather patterns making much of the planet uninhabitable (Monbiot 2007, 7–15; McPherson 2013).

Although New Catastrophist fears are grounded in scientific realities, and their proponents cannot necessarily perceive the ways in which their dire warnings might be misused, several elements of New Catastrophist theory on 'collapse' might be said to inadvertently evoke and provide context for the science fiction tropes of esoteric neo-fascists. Some New Catastrophist literature, for example, highlights how climate (and civilisational) collapse could potentially result in the 'Venusification' of Earth, creating Venus-like conditions that render the planet uninhabitable for all life forms (Hansen 2009, 224–236). Such visions risk being exploited by some accelerationist or esoteric neo-fascists, who find inspiration in Julius Evola's spiritual racism and the Nazi mythology of Aryan super-civilisations (Griffin 2007). This ideology asserts that the Aryan race represents the peak of human evolution and is responsible for all major accomplishments in science, arts and culture, but also that Aryan people come from an ancient civilisation and possess special spiritual powers (Kurlander 2017). While this 'race' maintained a hierarchical and elitist social structure, according to Evola, it was lost due to the impositions of an eschatological, decadent culture, and the decay of traditional societies. Other, fringe Nazi beliefs that existed in the context of WWII and even today posit the possibility of the extraterrestrial origins of the 'Aryan race' (e.g. Blavatsky 1888; Von Daniken 2018).

In the context of exacerbating the risks posed by climate change and the growing awareness of far-right responses to this, the required reductions in carbon emissions that correlate to avoiding eschatological, doomsday levels of warming and associated tipping points are also drastic. The IPCC targets propose emission reductions of at least 85% by 2050 (Spratt and Sutton 2008, 93). Many New Catastrophists agree with Hansen (2009, 140) that existing targets would only achieve CO_2 reductions to levels that geological records suggest are inadequate, and that we must aim for cuts of between 90% and 100% by 2030 in order to have a reasonable chance of avoiding dangerous tipping points and runaway, uncontrollable climate change (e.g. Hamilton 2010, 13; Gilding 2011, 128; Lynas 2011, 57). Currently the most ambitious post-2020 reduction pledges (that is, those not yet legislated) are in the order of 40–50% (adopted by China), with most major emitters (the US, EU, Japan) aiming for between 17% and 30% (Roelfsema et al. 2014, 794).

Although New Catastrophist activism and literature has transformative societal change in mind, it is also important to recognise that this literature is representative of an ostensibly pragmatic turn in environmental thought, which regards climate change as 'fundamentally a political problem' (Parenti 2011, 226). Its 'applied focus' (Barry 2012, 3) on 'workable solutions' (Spratt and Sutton 2008, 194) proposes decisive action in the political realm, presenting a critique of the ideological structures of the status quo in contemporary society, and calling for the stripping away of ideological illusions to enable sensible responses to material problems (Hamilton 2010, 197). Given its pragmatic focus, sometimes overlooked in the New Catastrophist literature is its inspiration from a radical environmentalism (Kovel 2007, 197; Hamilton 2010, 147) that has long had influential idealistic and utopian tendencies (Dobson 2007, 187–188, 202; Taylor 2008, 38–42). It is to these tendencies that the following part of this chapter turns, unpacking the barriers to change as diagnosed by different radical perspectives within New Catastrophist discourse, and examining the alternatives these perspectives provide to far-right climate change attitudes.

Economic barriers

New Catastrophist arguments about the barriers to necessary climate change mitigation include an emphasis on future civil conflict, the injurious impacts of over-consumptive societies, and the political and economic barriers to establishing post-capitalist economies. Despite the fact that these warnings are grounded in material reality and provide the basis for theorising progressivist 'utopian' visions, the diagnosis of these social ills that catalyse GHG emissions and harmful climate impacts is also in some ways amenable to misappropriation from the far right. In particular, the diagnosed ills according to New Catastrophist thinking can in some ways be said to hearken to far-right discourse on future impending 'race wars' (e.g. Saleam 1999), extreme-right proponents' criticism of 'decadent' and 'modern' liberal Western societies (Biehl 1995), and what some on the far and radical right in governments see as the necessity of removing national economies from international exchange, despite the deleterious impacts of protectionist measures on developing countries (Smith et al. 2023).

Drawing on both mainstream economics and Marxist arguments, many New Catastrophists, while accepting the basic need to curtail growth, argue that capitalism is a fundamental part of the current global economic order, and that any attempt to 'reform' capitalism in this regard without also overthrowing capitalist economic conditions generally is doomed to failure (e.g. Kovel 2007, 201–211; Foster et al. 2011, 201–206; Trainer 2011, 592). Aside from whether capitalism can be organised without growth or with a different concept of growth, or whether a socialist revolution is required or not, New Catastrophists tend to agree that, however it is described or achieved, serious changes need to be made to the way humans organise economically.

It is widely held by New Catastrophists, for example, that the economic 'status quo' based on exponential growth of production and consumption is incommensurate with meaningfully addressing climate challenge and meeting emissions reduction targets (e.g. McKibben 2007, 5–45; Speth 2008, 20, 52; McNall 2012, 40). New Catastrophism identifies serious difficulties in reaching emission reduction targets without imposing significant limitations on economic growth due to the global economic system's reliance on fossil fuels, which, regardless of efficiency gains, is continually increasing (Foster 2009, 124–128; McKibben 2007, 15–18). Also relevant is that while degrowth in developed countries would lead to a reduction in living standards and in over-consumption there, this would also yield significant deleterious impacts for residents in developing countries. Local economies and governance systems in Global South countries in particular are economically reliant upon wealthy states, especially given those states' historical globally extractivist resource and labour models of development.

The potential deleterious impacts of degrowth and no growth models in the Global North and wealthy countries for less wealthy states and the Global South are not really contested by degrowth think-tanks or UN agencies. These potential international effects are only the subject of an emergent literature, but New Catastrophists have for some time examined the difficulty of slowing or reversing economic growth due to the social effects generated domestically within both wealthy and less wealthy societies. In the current economic system, for example, a lack of growth equates to unemployment, social crisis and political instability (Smith 2010, 34; Lynas 2011, 240). It is also impossible in the current global political system for an elected leader to eschew growth (Spratt and Sutton 2008, xi–xii; Hamilton 2010, 49), and there is evidence that flagging growth can be a trigger for failing governments (MacCulloch 2001, 26, 27).

In the context of such effects, it is also important to acknowledge the downscale of economic activity required. New Catastrophists recognise Hamilton's (2010, 20) calculation that the IPCC's recommended 85% reductions by 2050 would require rich countries' emissions to contract by 6–7% per annum after a 2020 peak. However, if emissions do not peak until 2030, then reductions of 8.2% per annum will be required (Doniger et al. 2006, 764). Historical examples of emissions reductions in this approximate range have had severely negative impacts on economic growth. In this regard, many New Catastrophists point to the collapse of the Soviet Union, starting in the late 1980s, which saw CO_2 emissions reductions of around 5.2% per annum in the decade after the collapse and a halving of gross domestic product over the same period (Hamilton 2010, 20; Leahy et al. 2010, 854).

That the measures called for by the IPCC seem difficult to achieve is all the more troubling due to the New Catastrophist perception that they are inadequate for several reasons. And yet, the clear implications of even the most optimistic IPCC position are that deep, substantive systemic changes need to be implemented. The fact that this conservative mainstream body's analysis implies the inadequacy of simple reformist measures is one factor that reinforces the insistence on radical thinking in New Catastrophism. Due to the perceived intractability of dealing with climate change within an economic system predicated on continual exponential growth, there have been efforts to devise a theoretical economic order that does not rely on exponential economic growth, such as *Steady-state economics* (Daly 1977) and *Prosperity without growth* (Jackson 2009). Notwithstanding the afore-discussed issues, almost all New Catastrophists accept, and we agree, that growth must be addressed in order to meaningfully combat climate change, and many explicitly draw on ideal no-growth models (e.g. Shearman and Smith 2007, 130; Gilding 2011, 188; Klein 2014, 173).

The main disagreement among New Catastrophism proponents regarding no-growth economic models, then, comes down to whether such an economic change can be achieved without a larger revolution in the entire socioeconomic system. This is not always explicitly connected to the question of how degrowth impacts residents of LDCs and their economic and governmental systems; but the question may be implicit in New Catastrophist perspectives on the perceived necessity, or not, of revolutionary or transformative changes to the ways in which global human societies are organised.

Social and political barriers

Other social and political obstacles highlighted by New Catastrophists include the influence of political pressure groups stymieing climate change interventions. As this book has shown, the intersecting reactionary and conservative social interests of various neoliberal or 'economic rationalist' stakeholders converge within various institutions responsible for this, including lobbying firms like Lavoisier in Australia (see Chapter 4, p. 136). As many New Catastrophists point out, current fossil fuel reserves, if all were used, would create approximately five times the amount of carbon emissions required to 'tip' the climate into runaway change; therefore, the bulk of remaining fossil fuels can be said to be 'unburnable' (Hansen 2009, 173; Dalby 2013, 42–43). 'Leaving them in the ground' (Klein 2014, 153–154; McKibben 2012), though, contradicts the wishes of very powerful vested interests. The fossil fuel industry and fossil fuel states have fought back against the 'anti-carbon' conclusions of environmental science by attacking the science itself, funding an influential 'denial industry' to sow doubt in the public's mind regarding climate change and the need to implement reforms that would undermine the profitability of fossil fuels (Monbiot 2007, 20–42; Oreskes and Conway 2010).

Delay tactics have also latterly emerged, including 'carbon economy' corruption associated with carbon capture and storage technologies and renewable technology such as solar photovoltaics, the advocacy of which alongside the continued burning of fossil fuels and mining of rare earth minerals can distract from meaningful climate change mitigation (Bumpus and Liverman 2008; Martin and Walters 2013; see Chapter 4, p. 133).

Contrasting with right-wing resignatory responses to climate change that accept its iniquitous global effects, which were explored in the Australian context in some detail in Chapter 4, most New Catastrophists see the influence of powerful industrial entities on democratic political processes as a significant barrier to necessary change (Hansen 2009, 242; Hamilton 2010, 222–224; Klein 2014, 364). While this conclusion is reasonable and supported by evidence, resignation on the part of existing political institutions can risk contributing to far- and extreme-right actors' exploitation of the social politics and physical effects of global heating. The apparent inability of democratic political processes thus far to enact sufficient preventative measures has led to advocacy of authoritarianism and a rejection of formal (and informal) democracy on the right, also leading New Catastrophists typically from leftist positions to bemoan the 'paralysis' of liberal democracy due to its general commitment to the capitalist growth economy (Spratt and Sutton 2008, 214–216; Klein 2014, 360–361), its short-term focus (Cerutti 2008, 123; Orr 2009, 205), its adversarial character (Speth 2008, 219) and its susceptibility to corruption (Hansen 2009, 168, 224; Hamilton 2010, 223).

Another perceived political problem emphasised both by New Catastrophists and within the ESD literature is the increasingly common assumption that, as climate change proceeds, international cooperation will diminish and conflict will grow. As discussed in the account above, certain ESD securitising approaches to climate change bely the dominant historical contribution to climate change and right-wing white supremacy of global development through capitalist economies (see Richards 2023). As some of the New Catastrophist literature also emphasises, agonistic political scenarios of climate change devastation, in war, civil war, state failure, mass displacement and social unrest, which are widely assumed to accompany climate change and associated resource scarcity (Campbell et al. 2007; Dyer 2010, 7–25; Dalby 2014, 1), also contribute to asymmetric Global North–South violence. Many New Catastrophists assert such a vision of political turmoil and human suffering to be likely in the future under 'business-as-usual' approaches, and so they argue that this constitutes both a reason for revolution and a barrier to simple reforms (Kovel 2007, 16–17, 39; Foster 2009, 117–120; Parenti 2011, 13–20).

These arguments are in certain respects well substantiated, but it is important to recognise how rejections of representative democracy, capitalism and consumerism provide an evidentiary basis upon which far-right Romanticist and proto-fascist rejections of modernity and 'globalism' have been, albeit superficially, articulated. In some ways similar to the far right's rejection of Western decadence, many New Catastrophists also point to the fact that we, especially in the rich 'West', are addicted to a lavish, comfortable, consumerist lifestyle, and that the changes required to meet the climate challenge are widely perceived to be unattractive (Pérez and Esposito 2010). Many concede that it is particularly difficult to 'sell' a sustainable lifestyle defined by restraint, frugality and the like to those socialised into a culture based on opposing values (Monbiot 2007, 215; Hamilton 2010, 66–94). The historical tendencies towards communalist frugality on the political left provide a means of rejoinder to this; but so too does the contemporary right-wing emphasis on regressive, metapolitical means by which to destroy decadent liberal and, moreover, socialist traditions (Beiner 2018).

Various revolutionary perspectives on how capitalist consumerism is damaging the planet are further reinforced by the continually increasing global population, projected to reach 10 billion by 2100 (Dorling 2013, 1). Some New Catastrophists argue that the pursuit of a 'Western lifestyle' by growing populations worldwide would render any proposed energy-efficiency improvements, technological advancements or reforms to the current economic system insignificant (Spratt and Sutton 2008, xi; Dyer 2010, 48). In response to this issue, we would emphasise that New Catastrophist solutions often starkly contrast with rightist Malthusian approaches. Rather than endorsing parochial notions of national renewal, New Catastrophists tend to acknowledge the necessity for a global social revolution to accompany climate change action. This approach seeks to prevent the perpetuation of existing international disparities in wealth and lifestyle, and to avoid condemning the world's poorest populations to perpetual grinding poverty (Foster 2009, 117; Hillman et al. 2007, 190).

For most New Catastrophists, all of the above issues point to the apparent tragic inevitability of crisis. Through this light they lament the fact that, although we have known about the effects of our economic and social systems on the natural world for many decades, our reckless action continues (Shearman and Smith 2007, xii, xiv; Hamilton 2010, 207). Some see society as a 'social machine' adapted in a particular way and incapable of significant self-modification (Leahy et al. 2010, 857; Klein 2014, 63). This seeming inevitability is also then understood as having its own negative psychological effects, including fatalist apathy and simultaneous disbelief, which as this book has so far sought to show, can further reinforce inequalities and bring about ineffectual climate responses, rendering those susceptible vulnerable to recruitment by far-right social movements (Shearman and Smith 2007, 9; Marshall 2015, 6).

Whatever the reasons for the lack of effective climate action, New Catastrophists largely accept that some level of collapse is the most likely outcome of our predicament. Our progress towards the 'climate cliff' (Foster 2013, 1) is inexorable, and as 'crisis would be needed before society responds' (Gilding 2011, 126), crisis is therefore inevitable (Orr 2009, 193; Hamilton 2010, 132).

Degrowth, eco-socialism and eco-anarchism

In response to the impending climate crisis, New Catastrophist proposals for systemic change consist of visions of alternative social and economic systems, and political strategies for their realisation. Importantly, our critique is not intended to dismiss each of these or convey a false equivalence between radical left and radical right responses to the climate crisis. The tenets of New Catastrophist theory might be further developed into meaningful post-capitalist solutions, to which end we highlight the risks associated with some tenets of certain ideologies, in order to aid in this developmental process.

Differentiated New Catastrophist visions of a new society are usefully outlined by a New Catastrophist schema of possible future scenarios developed by the Stockholm Environment Institute's Global Scenario Group (GSG), *Great transition* (Raskin et al. 2002). *Great transition* lays out three broad categories of future scenarios, each with two possible iterations. 'Conventional Worlds' consists of policy reform and market forces. 'Barbarisation' consists of breakdown and fortress world. The eponymous 'Great Transition' scenarios are eco-communalism and the New Sustainable Paradigm (NSP). All of these scenarios are common in New Catastrophist narratives. The 'Conventional Worlds' scenarios – market forces and policy reform, which propose no fundamental societal alteration – are seen by New Catastrophists as the source of the problem, and proposals based on these positions are therefore understood to be grossly insufficient (e.g. Foster 2009, 260; Klein 2014, 64–83).

Under the differentiated New Catastrophist framing of the crisis, outlined in the previous section, it is precisely due to the failure of market forces and of attempts at policy reform to successfully address climate change that the status quo is heading towards 'Barbarisation' – at worst breakdown, with compounding out-of-control crises, 'unbridled conflict', and social and economic collapse; and at best the fortress world of widespread collapse and authoritarian defensive reaction from elites isolated in securitised enclaves (Foster 2009, 117; Parenti 2011, 10, 223; Klein 2014, 59). The eco-communalism scenario should be recognisable as the default ecotopian position of radical greens. It incorporates the familiar green vision of 'bio-regionalism, localism, face-to-face democracy, small technology and economic autarky' (Raskin et al. 2002, 15), alongside a no-growth economy with production for need, not profit (O'Riordan 1981, 307; Martell 1994, 151). Virtually all New Catastrophists share some level of commitment to this democratic, no-growth, localist model (Foster 2009, 264; McKibben 2007, 103; Levene et al. 2010, 79; Klein 2014, 364–365).[1]

The prevalence of localist visions reflects the historically strong 'anti-statist' current in radical environmentalism (Kovel 2007, 197; Taylor 2008, 46). Although there have recently been concerted efforts to address the perceived weaknesses of this anarchistic 'pessimism' (Barry and Eckersley 2005, x) regarding a positive role for the state in environmental protection (see Eckersley 2004), this tendency retains a strong influence in radical green thinking, including in many New Catastrophist proposals. While eco-anarchist New Catastrophists continue to see this small, local vision – or small-seed communism – as the *entire* shape of a future society (Gelderloos 2010; Trainer 2011, 80), others, while accepting localism as an essential element, presume the need for others, especially the state, to ensure that such local efforts have a chance of emerging and surviving (e.g. Kovel 2007, 197; Parenti 2011, 242; Klein 2014, 404–405). So, while there might appear to be much dividing anarchist and statist New Catastrophist proposals, and it may seem contradictory for 'top-down' statist proposals to include this localist vision, there is in fact no fundamental disagreement among proponents of New Catastrophism that the ideal localist model be an essential element of a New Catastrophist ecotopia.

By the GSG's own admission, its vision of a 'new paradigm' is not very different from this familiar green vision of ecotopia, but it seeks to change rather than abandon the urban and industrial model, and to renew globalism rather than 'retreat into localism' (Raskin et al. 2002, 16). NSP, more explicitly than eco-communalism, retains an important role for the state and international organisations as facilitators of an environmentally conscious international regime, and a 'global network of localisms' (Raskin et al. 2002, 44–45; Chesters and Welsh 2006, 161, 163). Rather than an alternative, this merely represents a more detailed account of the standard ecotopian vision. Practically all New Catastrophists envision a similar scenario as their proposal for a new model of social organisation, which falls somewhere on the spectrum between the vision of eco-communalism and NSP, primarily differing in how they hope to achieve it, and what other elements they see as necessary to support the localist vision.

A revolution by any other name

Few New Catastrophists have the commitment to call their intended social change 'revolution'. But regardless of whether it is labelled a 'transition', 'pivot' or 'paradigm shift'; led from the 'top down' or the 'bottom up'; or by a 'mass movement' or an 'enlightened technocracy', what New Catastrophism theorists advocate, according to the scope and depth of systemic change envisioned, is global social revolution. Movements that aim to alter fundamental aspects of a society can properly be considered revolutionary, regardless of how they self-describe, at the very least due to the obstacles they face (DeFronzo 2011, 8). Some 'primitivist' New Catastrophists amenable to co-optation by right-wing neo-Malthusian arguments see the breakdown scenario as itself the revolution that is needed, and advocate embracing civilisational collapse, depopulation and a return to a simpler, sustainable 'new tribalism' that could be humanity's 'next great adventure' (Farnish 2009; Quinn 2000; Homer-Dixon 2010). However, most New Catastrophists ostensibly aim at preventing complete collapse, their methods broadly dividing into 'top-down' and 'bottom-up' approaches, with most proposals being some combination of both. The degree to which a proposal sees a positive role for the state generally determines the weighting towards top-down rather that bottom-up approaches, and this usually corresponds with whether the new society aimed for lies closer to the eco-communalism (non-state) or NSP (state) scenario.

Bottom-up methods include calls for individuals to 'take action to reduce emissions' and 'buy green' (Flannery 2006, 254; Hillman et al. 2007, 241–243). They also feature efforts at 'prefiguring' the new society 'in the shell of the old' (Graeber 2004, 7) by building local, and community models of societal organisation and governance intended for general application (Kovel 2007, 207–211; Trainer 2012, 597). Otherwise, they entail tactics of nonviolent revolution, including mass protest, civil disobedience and direct action against ecologically harmful practices (such as fossil fuel extraction) (Hamilton 2010, 225; Klein 2014, 139–140). Examples of a purely bottom-up approach include explicitly eco-anarchist plans for a stateless variety of the eco-communalism scenario brought about by prefiguration, voluntary disassociation (that is, 'choosing a Simpler Way') and devolution of power to a local level (Trainer 2012, 598), or by these methods in combination with violent insurrection (Gelderloos 2007, 89, 114).

Some eco-socialist visions are quite similar to eco-anarchism, employing mainly bottom-up methods, such as prefiguration and nonviolent tactics to challenge the status quo.[2] But rather than abolish the state, they aim to seize and transform it into a revolutionary force via a mass coalition of 'new social movements', and eventually election (Kovel 2007, 258); or by classic Marxist proletarian organisation and collective action occurring initially in the Global South (Foster et al. 2011, 439).[3] Eco-socialists, like most New Catastrophists, propose a combination of top-down and bottom-up approaches characterised by a mass movement forcing politicians to act through drastic policy measures (Monbiot 2007, 214; Gilding 2011, 263; Klein 2014, 352), like shutting coal power stations (Flannery 2006, 249; Hansen 2009, 241), large-scale state investment in green technology (Parenti 2011, 239) and rationing essential resources (Cox 2013).

In contrast to eco-socialist approaches, other, demagogic and authoritarian responses to the climate crisis within the New Catastrophist literature are more directly redolent of technocratic, elitist and other authoritarian responses also present in securitising approaches to the ESD, and they carry greater risks of entrenching regional (and perhaps, ethnonationalist) divides in respect of climate change and its social consequences. Those in question directly advocate a purely 'top-down' revolution, having lost all faith in the ability of 'grassroots' efforts to seize power, or even to provide the direction or the pressure required to implement the necessary change in the time available (Lynas 2005, 3; Parenti 2011, 241). This approach varies between radical (that is, 'non-reformist')[4] reforms implemented from above (Campbell 2009, 15–17; Lynas 2011, 228, 243), sudden drastic 'emergency measures' by the state apparatus (Dyer 2010, 197–199; Gilding 2011, 119; Cox 2013, 107, 260) and seizure of the state by an enlightened technocratic 'eco-elite' (Shearman and Smith 2007, 141).

The first path requires a spontaneous recognition of the gravity of the situation by political and business leaders. Here the system changes itself; it is not 'regime change' but a surge in elite 'political will' that instigates revolution (Hillman et al. 2007, 193). The 'state of emergency' option envisions a drastic response by current regimes enabled by a suspension of democracy and/or individual liberties (Cox 2013, 20, 82), or in the absence of such moves, power seizure by a technocratic elite (Shearman and Smith 2007, 15, 131–134). Both paths explicitly rely on the issue being forced by the onset of crisis. A popular analogy for this is the response of allied powers to the production challenges of WWII (e.g. Cox 2013, 22–41). These proposals see change coming through a war-economy response to the onset of crisis. Under a 'state of emergency', action will be taken to usher in sudden, sweeping changes that become permanent, with further change to continue as the initial state of emergency is lifted (Spratt and Sutton 2008, 227, 247–250; Gilding 2011, 110, 253).

While 'war-effort' proponents and 'primitivists' explicitly argue that collapse is the catalyst that will spur change, ostensibly less extreme New Catastrophist proposals also logically entail 'revolution via collapse' in their revolutionary theory. All New Catastrophists perceive a seemingly inexorable acceleration towards crisis, and all seek a catalyst to alter this dire trajectory and make progress towards radical change possible. New Catastrophist thought as a whole tends to see some crisis, or partial collapse, as an essential trigger for 'awakening' us to the reality of climate change, thus sharing some characteristics with radical climate responses articulated on the revolutionary and insurrectionist right. This similarity is even figuratively demonstrated in popular far-right phrases that refer to a sudden dawning of recognition about the topsy-turvy nature of the world, which leads to ideological acceptance of fascism, such as the notion of 'awake not woke' (e.g. Mering 2021), or the QAnon notion of 'the Great Awakening' (Amarasingam and Argentino 2020; Busbridge et al. 2020).[5]

The potentially accelerationist aspects of some New Catastrophist proposals therefore pertain mostly to their common acknowledgement that a catalyst is required to make the proposed changes possible. Although such proposals are usually framed in terms of collapse prevention, proponents of bottom-up efforts such as prefiguration argue that 'only serious scarcity will jolt people into action' and local community-building efforts can at most hope to 'influence' the shape of post-collapse society (Trainer 2012, 598). Those calling for mass movements propose that they must wait until 'another crisis hits' to 'get us in the streets and squares once again' and have any hope of influencing the course of events (Klein 2014, 466). Eco-socialists claim that the 'ecology of revolutionary change' (Foster 2009, 277) will increase the viability of paths for social change which are currently impossible, and that it is 'only a matter of time' before the situation becomes one of 'explosive urgency' (Kovel 2007, 242). Civil disobedience advocates see such actions as 'slow[ing] down … the effects' and 'preparing for the impact', even though 'it is too late to prevent climate disruption' (Hamilton 2010, 222, 223). And proponents of revolutionary state seizure, whether of the left or right, believe that the 'revolutionary situation' (Tilly 1978, 189) depends on the inability of either liberal regimes or capitalism to 'deal with crisis' (Foster 2013, 17–18; Shearman and Smith 2007, xiii).

The risks of consequentialism

The various disparate accelerationist, populationist or authoritarian elements that occupy the overlap of New Catastrophist theory with far-right climate change responses relate to the advent of New Catastrophist 'consequentialism'. Consequentialism consists of the belief that the consequences of one's actions should be the basis for judging the morality of those actions, while in political terms, it is often summarised by the assertion that 'ends justify means' (O'Boyle 2002, 25). Revolutionary consequentialism signifies the conviction that any and all actions that lead to the desired radical social transformation are not only justified, but morally good (Mayerfield 1999, 123; O'Boyle 2002, 27). The threat of extinction is perhaps the best imaginable motivation for revolutionary consequentialism in respect of global heating, with some New Catastrophists perhaps predictably showing signs of succumbing to this pressure.

There has long been a revolutionary consequentialist fringe of extreme environmentalists influential on the contemporary far and extreme right since Thomas Malthus, including Dave Foreman and Pentti Linkola. These figures have applauded famine and disease for its presumably positive environmental consequences (Foreman in Devall 1986), and have argued that mass depopulation through war, genocide and sterilisation is morally justified to prevent climate catastrophe (Linkola 2011, 131–133).[6] Although New Catastrophists at the time of this book's writing do not widely advocate such extreme measures, the seemingly insurmountable obstacles to implementing what they see as non-negotiable change have led to statements of desperation that, we suggest, can display a drift towards revolutionary consequentialism.

Although many New Catastrophists do intend to raise public political consciousness in order to bring about effective mitigatory and transformative climate responses, the inflexible measures they sometimes advocate can risk playing into 'collapse' ideologies that sanction catastrophic humanitarian outcomes and produce an ideological milieu that provides a supportive context for the far right. For instance, some New Catastrophists argue that 'the stakes are too high to settle for anything else' (Klein 2014, 466) and 'time is too short' (Foster 2013, 18), that we can evolve 'to a higher evolutionary state … because we have to' (Gilding 2011, 4), and that we need to bring about social and ecological revolution 'no matter how devastating the consequences may be' (Fanon, cited in Klein 2014, 459). Combined with a misguided belief in the New Catastrophists' ability to discern with certainty a correct path, this attitude – albeit through different forms of radical, anarchist, eco-Marxist or corporate green expressions – represents a dangerous perspective to take into a highly charged future ideological battle over humanity's survival.

While a communitarian view of human nature underpins many New Catastrophist proposals, those few who do not endorse the democratic, no-growth local model make similarly idealistic assumptions. For instance, Lynas (2011), one of the only New Catastrophists to explicitly reject the 'steady-state' economic models of Daly (1977) and Jackson (2009), falls back on the typical economic assumptions associated with liberal capitalism (Lynas 2011, 242). And perhaps the only New Catastrophists to reject democracy in favour of authoritarian rule characterise humanity as essentially obedient (Shearman and Smith 2007, 99–103), and yet still manage to see the localist, no-growth vision (minus the direct-democracy aspect) as a fundamental element of their enlightened authoritarian eco-society (Shearman and Smith 2007, 84).

As recognised by progressivist New Catastrophists, but also on the right implicitly in the various currents of (albeit often anti-intellectual) ecofascist and other accelerationist media (Loadenthal 2022), a successful revolution requires the existence (or creation) of a 'revolutionary situation', and a revolutionary cadre or coalition with the means to transform this potential into an actual 'revolutionary outcome' (Tilly 1978, 189, 195). However, 'revolutionary reality is complex' (Tilly 1978, 189) and 'it is not every revolutionary situation that leads to revolution' (Lenin 1980, 213). Revolutions are unpredictable, chaotic, often violent, and often come with significant negative outcomes. Failed left revolutions can spur the rise of fascism (Benjamin, cited in Žižek and Park 2013, 25) and have been the 'catalyst for some of the most prolonged and bloody wars' (Conge 2000, 133). The ongoing tragic spiral of violence and destruction that has engulfed Syria (and surrounding regions) after its failed revolution of 2011 offers a stark modern example of these revolutionary risks (Holliday 2011; Gelderloos 2013; Davies 2014, 312–313).

Despite the acceptance that the Global South will face the most extreme change in material conditions as a result of climate change (and so, presumably, the greatest motivation for social change) (Dyer 2010, 18–21; Parenti 2011, 8–9), it is also thought by most New Catastrophists that given wealthy states' contributions to climate change, the eco-revolution must occur mainly, or most urgently, in the rich North (Spratt and Sutton 2008, 138; Foster et al. 2011, 440). How pressure for change in the South is to translate into real action in the less impacted North is not explained. The most explicit attempts to address this jump from abstract national ideal models to actual global social and political change generally get no more specific than asserting that revolutions in the South be '*somehow* mirrored … in the advanced capitalist world' (our italics) (Foster 2009, 276). Worldwide revolutionary waves, or cascades, are rare, slow and 'often accompanied by overt foreign invasion' (Walt 2011). On the rare occasion that events in the developing world *do* inspire responses in the developed world, decidedly different local conditions underlie responses to the same issue and therefore influence the potential for effective change. The 'Occupy' movement – the Western answer to the Arab Spring – demonstrates, in its comparatively limited capacity to bring about long-term change, the difficulty of transferring a movement for change that is grounded in real material difficulty and corrupt authoritarian rule to the developed world, even when both movements are ostensibly opposing local iterations of the same global system (Mason 2012, 184–188; Graeber 2013, 17).

Our current predictions of the climate future are also largely linear extrapolations of processes already underway and can only give us a vague idea of the general context within which more drastic, unpredictable change is likely to occur. Climatic conditions and changing weather patterns, attributed to the recent warming level of 1.15°C above PIA (WMO 2022, 3), are already negatively affecting many millions of people, with material hardship and a daily struggle for survival the current reality for much of the world's population (Parenti 2011, 9–11). This situation is becoming more acute as a result of the climate change we have already experienced, and any sudden shock or tipping point is likely to plunge many more people into such precarious conditions (Monbiot 2007, 8; Dyer 2010, 18–21). Therefore, revolution via collapse will often mean material hardship to the point of a threat to survival. Frequently, of course, the threat of collapse will represent a challenge that cannot be met; many will die to provide the catalyst for the serious change on which New Catastrophist proposals rely.

Despite these dire realities, some New Catastrophists go further than believing ecotopian revolution possible under the assumed conditions of global climate and social crisis, and imply that positive outcomes are practically inevitable, thus appearing optimistic about the prospect of impending catastrophe. Naomi Klein, for example, argues that 'climate change is our chance' to undertake 'the unfinished business of liberation' (that is, social and economic justice) (Klein 2014, 59, 459). Rebecca Solnit emphasises that natural disasters inspire 'pro-social' behaviour, and as climate change is set to deliver ever-increasing natural disasters, it is a 'chance for decentralization, democratization, civic engagement, and emergent organizations' (Solnit 2010, 8). Klein and Solnit's optimism could in our view be interpreted as potentially perilous examples of 'apocalyptic opportunism' or 'disaster utopianism' (Walker 2010), a state that is certainly not unique to them; the basic message of salvation through apocalyptic rebirth is practically ubiquitous in New Catastrophist communications.

Finally, there are those within New Catastrophism who believe that capitalism can and must reform itself to internalise environmental costs or accept a no-growth norm. These figures also perhaps overly optimistically see the crisis as a 'once-in-all-lifetimes opportunity' to create 'a more just and peaceful world in which extreme poverty does not exist and universal rights and freedoms are extended to all of humanity' (Nadeau 2006, 186–187), and an 'exciting and uplifting ... "once in a civilisation" opportunity' to make a difference greater than any in human history (Gilding 2011, 5, 180). Eco-socialists have even cast this 'fantastic challenge' as a 'chance to transform the relationship between humanity and nature' and 'build a better world ... beyond capital' (Kovel 2007, 10, 279), as well as a justification for state socialism and 'genuine socialist planning' (Foster 2009, 275–276). Authoritarians advocate their own version of utopia, believing that the climate crisis can trigger a 'new Enlightenment' that will lead to an 'unsullied system' (Shearman and Smith 2007, 85, 165–167).

Concluding comments

In response to some elements of New Catastrophist and ESD projections, we would emphasise that to suggest that we can ride the apocalypse in the 'sweet spot' between the onset of collapse and the point at which its completion becomes inevitable is misguided. We have no way of knowing whether by the time the crisis is severe enough to spur mass action, there will be any, let alone sufficient, time to enact the requirements of a specific plan. We must admit that the limits of our knowledge are insufficient to confidently predict, and therefore confidently plan, any particular response to the crisis, as we do not know how it will unfold. Perhaps more importantly, we cannot with confidence dismiss the possible applicability of other paths (Haynes 2003, 36, 38), as New Catastrophists often do in their advocacy of various versions of 'the only way'. Even if it does somehow work out to be possible, presuming the ability to manage massive, complex and interrelated systems such as the global ecosystem, its climate, and the human social world in our view displays a serious overconfidence in humanity's abilities.

A critical assessment of this state of affairs also requires acknowledging that those with hopes of creating an ecotopia based on a singular vision and certain path downplay the reality of global revolution under complex crisis conditions. Among other problems with this approach is that it risks devaluing the lives and potential suffering of vast swathes of humanity. This standpoint also frequently maintains the privileged position of the West, especially the intellectual environmentalist and social visionary elite, as the architects of a new, sustainable, socially just society based on 'universal' ecotopian values, which in reality primarily reflect Western perspectives, concerns and, importantly, leadership in the transition to a new world.

As this chapter has sought to show, both the New Catastrophist and ESD literatures represent the main influences – academic and public alike – on the dominant popular narratives and conceptual context within which far-right environmental responses develop. They can also normalise a narrative and zero-sum logic that inadvertently reinforces several tropes and fantasies of the far and extreme right. The unwavering certainty of some New Catastrophist arguments – their ubiquitous promotion of solutions claimed as the only option for survival – normalises an absolutist stance of 'my way or extinction', which might seem defensible when your 'way' is radical localised democracy but may be more dangerous when employed in the service of an alternative strategy or social vision. The focus on population, displacement and immigration as major threats emerging from climate change, in New Catastrophist but especially in ESD discourses, can also feed into far-right Great Replacement narratives of demographic displacement and non-white immigration into historically white-dominant societies.

Many of these issues are of critical importance to the Australian context. As the driest inhabited continent, Australia is, of course, at the forefront of many climate effects; and as an island nation, it has had significant advantages in establishing a securitised immigration system that demonises and disincentivises refugees. As discussed in Chapter 2, the far right in Australia rests on a longstanding base of institutional white supremacy and fears of demographic replacement of the white majority, which in part inspired the country's foundation as an independent nation-state.

Australia is also founded on an extractivist mentality born of the early colonisers' desire and ability to exert control and ownership of the land, and this orientation is apparent still in its contemporary economic reliance on resource exploitation. As detailed in Chapters 2 and 4, resource (in particular fossil fuel) extractivism in colonial Australia was driven by cultural origin narratives of dominating the environment, combined with capitalist accumulation incentives and the desire to extract profits from the land. Settlers viewed the land as a resource to be exploited for economic gain, rather than as a shared resource requiring stewardship and care, or as a coexistent entity within the biosphere.

While the extraction and use of natural resources in Australia and globally have made possible the technological developments, systems and infrastructure many of us now benefit from and rely upon, the unfettered use of natural resources motivated by capital accumulation and a lack of concern for the natural environment have also led to devastating social and environmental outcomes. Legacies of dispossession and exploitation in Australia have contributed to the development of our current dominant culture in which the unyielding and unsustainable exploitation of natural resources is constructed as a natural and inevitable part of our national economy (Harvey 1996). Australia does of course have a history of significant contributions to environmentalism, from establishing one of the world's first dedicated 'green' political parties in the 1972 United Tasmania Group, to developing the alternative food production system 'permaculture'. Yet it has also been a breeding ground for international acts of white supremacist political violence, including those inspired by the stated environmental priorities of the violent white supremacist actor who perpetrated a mass shooting at two mosques in Christchurch in 2019.

As highlighted at various stages in this book, we do not claim to have an easy or obvious solution to the immense challenge posed by global heating, nor do we as social scientists have the expertise to theorise practical solutions. We do, however, suggest that based on the empirical and conceptual insights set out in this book, it should be uncontroversial to agree with New Catastrophists that the continuation and acceleration of the current system of relatively unconstrained extraction, exploitation, consumption and destruction appear to be leading us towards catastrophic climate outcomes. Common New Catastrophist and ESD narratives featuring monistic and dogmatic assertions of singular solutions, confidently presented as the inarguable path to salvation, are, however, problematic in many ways. The most dangerous potential consequences of these narratives render them vulnerable to far-right co-optation, including on a micro scale in acts of non-state political violence, and on a macro level through structural violence enacted by states and major institutions, exemplified by the processes of securitisation, containment and exclusion.

The possible violent outcomes of institutional and radical visions claiming to be the only viable means of addressing the growth of the far right and extinction-level threats from climate change are not the only potential undesirable outcomes associated with such approaches. As this chapter has explored, another problem with both ESD and New Catastrophist approaches may be the conceited provision of environmental, social and political solutions by individuals writing from an elite position, who may be disconnected from environmental and social movements, and from the grassroots level of sustainable environmental management. This political positioning is especially troublesome in settler-colonial societies built on the destruction of longstanding, pre-existing Indigenous systems of complex eco-management with demonstrated sustainable longevity, as seen in Australia.

Prior to colonial invasion, the landmass now known as Australia was intricately managed with a degree of complexity and detail almost inconceivable in today's world, often baffling early settlers. Some early observers at the time of settlement recognised the landscape not as untamed wilderness but as a highly regulated, complex system maintained by patient and deliberate application of human labour over countless generations. They described what existed before the destruction by colonial forces as 'the biggest estate on Earth', as it appeared to European eyes to be reminiscent of vast, carefully manicured gardens and hunting grounds cultivated to provide for human needs while maintaining ecological sustainability (Gammage 2011). This was arguably the world's largest sustainably managed ecosystem existing at such a level of detail, and achieved through diligent and conscious development over more than 65,000 years by the hundreds of Indigenous nations across the continent.

In such a context, any attempt to devise novel solutions from a higher (Western) authority, which aim to replace the current system of rapacious exploitation without considering that the current system displaced relatively recent successful examples of large-scale sustainable environmental management, can appear both hubristic and short-sighted. This is not to suggest that the solution to climate change lies merely in returning to a pre-industrial past led by Indigenous cultures. Unfortunately, the global capitalist system of environmental exploitation has generated wide-ranging and deep-seated problems that are not resolvable at the local level and may sometimes exceed the remedial capacity of traditional approaches, such as global industrial atmospheric and oceanic pollution. Moreover, the global human population has increased substantially over many millennia and can likely now only be sustained by new technologies beyond agrarian living. These factors do not mean, however, that traditional Indigenous approaches to sustainability lack contemporary significance.

Most of the modern industrial world is built on Indigenous lands, where ecological knowledge and sustainable systems of environmental management have been devalued and displaced but not completely destroyed. When searching for alternative models of social and environmental organisation, rather than attempting to generate entirely novel systems to impose anew on stolen lands, it might be reasonable to foreground the perspectives of those who, until very recently, had achieved sustainable human–nature relationships upon which our survival as a species likely depends.

Notes

1 For the sake of brevity, we refer to this familiar green ideal social vision as 'localism', though we recognise that this is an oversimplification. Because this ecotopian vision is such a strong tendency in environmental thought, we presume the reader will not have trouble calling it to mind, and will accept this label as referring to the more comprehensive description just given.
2 Prefiguration seeks to supplant the dominant socioeconomic system by good example and eventual peaceful replacement. Examples of this tactic include intentional communities or 'communes', local economic schemes such as local currencies and community-supported agriculture (that is, collectively owned and/or subscriber-based local food production), and encampments such as the 'Occupy' movement (see Graeber 2002, 72; Kovel 2007, 207).
3 Here we employ the terms 'North' and 'West' interchangeably to refer to more developed economies, and the terms 'South' and 'non-West' to refer to 'developing', often post-colonial, economies. This follows common practice in the New Catastrophist literature (see Hillman et al. 2007, 186).
4 'Non-reformist' reforms, a phrase coined by Andre Gorz, refer to measures that are intended to, in the long term, have a radical effect on fundamental aspects of society yet

are imagined to come about through the application of gradual 'reforms' (see Gorz 1967, 7; Hobsbawm 1977, 21).

5 The QAnon conspiracy contends that a secretive international network of Satanic child exploitation is controlled by Jewish global(ist) political and commercial elites. Proponents argue that the Gates Foundation used COVID-19 vaccinations to microchip people for mass surveillance (Busbridge et al. 2020).

6 Foreman subsequently retracted his most controversial statements, though Linkola has not retracted his (see Foreman 1991, 107–108).

References

@antonioguterres (5 April 2022) 'Climate activists are sometimes depicted as dangerous radicals' [Twitter], accessed 17 November 2022. https://twitter.com/antonioguterres/status/1511294073474367488?lang=en

Abrahams D (2019) 'From discourse to policy: US policy communities' perceptions of and approaches to climate change and security', *Conflict, Security & Development*, 19(4): 323–345.

Ahmed N (2022) 'UN warns of "total societal collapse" due to breaching planetary boundaries', *Byline Times*, accessed 1 December 2022. https://bylinetimes.com/2022/05/26/un-warns-of-total-societal-collapse-due-to-breaching-of-planetary-boundaries/

Amarasingam A andArgentino MA (2020) 'The QAnon conspiracy theory: a security threat in the making', *CTC Sentinel*, 13(7): 37–44.

Anson A (2022) *Against the eco-fascist creep: debunking ecofascist myths*, ASLE, accessed 1 January 2023. https://www.asle.org/teaching_resources/against-the-ecofascist-creep/

Baer HA (2012) *Global capitalism and climate change: the need for an alternative world system*, AltaMira Press, Virginia.

Barnett J (2019) 'Global environmental change I: climate resilient peace?', *Progress in Human Geography*, 43(5): 927–936.

Barry J (2012) *The politics of actually existing unsustainability: human flourishing in a climate-changed, carbon constrained world*, Oxford University Press, Oxford.

Barry J and Eckersley R (2005) *The state and the global ecological crisis*, MIT Press, Cambridge.

Beiner R (2018) *Dangerous minds: Nietzsche, Heidegger, and the return of the far right*, University of Pennsylvania Press, Philadelphia.

Berglund O and Schmidt D (2020) *Extinction rebellion and climate change activism: breaking the law to change the world*, Springer Nature, London.

Biehl J (1995) '"Ecology" and the modernization of fascism in the German ultra-right', *Society and Nature*, 2(2): 130–170.

Blavatsky HP (1888) *The secret doctrine*, Books on Demand, Norderstedt.

Bloom P (2016) *Beyond power and resistance: politics at the radical limits*, Rowman & Littlefield, Lanham.

Brinn G (2022) 'The path down to green liberalism', *Environmental Politics*, 31(4): 643–662.

Bumpus AG and Liverman DM (2008) 'Accumulation by decarbonization and the governance of carbon offsets', *Economic Geography*, 84(2): 127–155.

Busbridge R, Moffitt B and Thorburn J (2020) 'Cultural Marxism: far-right conspiracy theory in Australia's culture wars', *Social Identities*, 26(6): 722–738.

Busby J (2018) 'Taking stock: the field of climate and security', *Current Climate Change Reports*, 4(4): 338–346.

Campbell KM (2009) *Climatic cataclysm: the foreign policy and national security implications of climate change*, Brookings Institution Press, Washington.

Campbell KM, Gulledge J, McNeill JR, Podesta J, Ogden P, Fuerth L, Woolsey RJ, Lennon AT, Smith J and Weitz R (2007) *The age of consequences: the foreign policy and national security implications of global climate change*, Center for Strategic and International Studies, Washington.

Cerutti F (2008) *Global challenges for Leviathan: a political philosophy of nuclear weapons and global warming*, Lexington Books, New York.

Chesters G and Welsh I (2006) *Complexity and social movements: multitudes at the edge of chaos*, Routledge, New York.

Childs C (2013) *Apocalyptic planet: field guide to the future of the earth*, Vintage, New York.

Conge PJ (2000) *From revolution to war: state relations in a world of change*, University of Michigan Press, Ann Arbor.

Cox S (2013) *Any way you slice it: the past, present, and future of rationing*, The New Press, New York.

Dalby S (2013) 'The geopolitics of climate change', *Political Geography*, 37: 38–47. https://doi.org/10.1016/j.polgeo.2013.09.004

Dalby S (2014) 'Rethinking geopolitics: climate security in the Anthropocene', *Global Policy*, 5(1). https://doi.org/10.1111/1758-5899.12074

Daly HE (1977) *Steady-state economics*, W H Freeman, San Francisco.

Davies TR (2014) 'The failure of strategic nonviolent action in Bahrain, Egypt, Libya and Syria: "political ju-jitsu" in reverse', *Global Change, Peace & Security*, 26(3): 299–313.

Davis DH (2007) *Ignoring the apocalypse: why planning to prevent environmental catastrophe goes astray*, Greenwood Publishing Group, Westport.

DeFronzo J (2011) *Revolutions and revolutionary movements*, Westview Press, Boulder.

Detraz N and Betsill MM (2009) 'Climate change and environmental security: for whom the discourse shifts', *International Studies Perspectives*, 10(3): 303–320.

Devall B (1986) 'Dave Foreman interview', *Simple Living*, 2(12): 2–4.

Diamond J (2005) *Collapse: how societies choose to fail or succeed*, Penguin Books, London.

Dobson A (2007) *Green political thought*, Routledge, London.

Doniger DD, Herzog AV and Lashof, DA (2006) 'An ambitious, centrist approach to global warming legislation', *Science*, 314: 764–765.

Dorling D (2013) *Population 10 billion*, Hachette, London.

Dyer G (2010) *Climate wars: the fight for survival as the world overheats*, Simon and Schuster, New York.

Eckersley R (2004) *The green state: rethinking democracy and sovereignty*, MIT Press, Cambridge.

Eco U (1995) 'Ur-fascism', *The New York Review of Books*, 42(11): 12–15.

Ehrlich PR (1968) *The population bomb*, Ballantine Books, New York.

Farnish K (2009) *Time's up!: an uncivilized solution to a global crisis*, Green Books, Cambridge.

Flannery T (2006) *The weather makers: how man is changing the climate and what it means for life on earth*, Grove Press, New York.

Floyd R (2015) 'Global climate security governance: a case of institutional and ideational fragmentation', *Conflict, Security & Development*, 15(2): 119–146. https://doi.org/10.1080/14678802.2015.1034452

Forchtner B (ed) (2019) *The far right and the environment: politics, discourse and communication*, Routledge, London.

Foreman D (1991) *Defending the earth: debate between Murray Bookchin and Dave Foreman / foreword by David Levine*, Black Rose Books, Montreal.

Foster JB (2009) *The ecological revolution: making peace with the planet*, Monthly Review Press, New York.

Foster JB (2013) 'James Hansen and the climate-change exit strategy', *Monthly Review*, 64(9): 9–28.

Foster JB, Clark B and York R (2011) *The ecological rift: capitalism's war on the earth*, NYU Press, New York.

Gammage B (2011) *The biggest estate on earth: how Aborigines made Australia*, Allen and Unwin, Crows Nest.

Garzón A (29 April 2022) 'The limits to growth: eco-socialism or barbarism', *La-U*, accessed 1 June 2022. https://la-u.org/the-limits-to-growth-eco-socialism-or-barbarism/

Gelderloos P (2007) *How nonviolence protects the state*, South End Press, Boston.

Gelderloos P (2010) 'An anarchist solution to global warming', *Guerilla News*, accessed 18 December 2014. https://guerrillanews.wordpress.com/2010/09/16/an-anarchist-solution-to-global-warming/

Gelderloos P (2013) *The failure of nonviolence*, Left Bank Books, St Louis.

Gilding P (2011) *The great disruption: how the climate crisis will transform the global economy*, Bloomsbury Publishing, London.

Gorz A (1967) *Strategy for labor: a radical proposal*, Beacon Press, Boston.

Gorz A (1980) *Ecology as politics*, Black Rose Books, Montreal.

Gow K (2009) *Meltdown: climate change, natural disasters, and other catastrophes – Fears and concerns of the future*, Nova Science Publishers, New York.

Graeber D (2002) 'The new anarchists', *New Left Review*, 13(1): 61–73.

Graeber D (2004) *Fragments of an anarchist anthropology*, Prickly Paradigm Press, Chicago.

Graeber D (2013) *The democracy project: a history, a crisis, a movement*, Random House, New York.

Griffin R (2007) *Modernism and fascism: the sense of a beginning under Mussolini and Hitler*, Palgrave Macmillan, London.

Griffin R (2018) *Fascism*, John Wiley & Sons, New Jersey.

Hamilton C (2010) *Requiem for a species: why we resist the truth about climate change*, Allen & Unwin, Crows Nest.

Hansen J (2007) 'Scientific reticence and sea level rise', *Environmental Research Letters*, 2(2): 1–6.

Hansen J (2009) *Storms of my grandchildren: the truth about the coming climate catastrophe and our last chance to save humanity*, Bloomsbury Publishing, New York.

Hansen JE, Sato M, Simons L, Nazarenko LS, von Schuckmann K, Loeb NG, Osman MB, Kharecha P, Jin Q and Tselioudis G (2022) 'Global warming in the pipeline', *arXiv*. https://doi.org/10.48550/arXiv.2212.04474

Hartmann T (2013) *The last hours of humanity*, Waterside Productions, Cardiff.

Haynes P (2003) *Managing complexity in the public services*, McGraw-Hill International, New York.

Harvey D (1996) *Justice, nature and the geography of difference*, Blackwell, London.

Heinberg R (2005) *The party's over: oil, war and the fate of industrial societies*, New Society Publishers, Gabriola Island.

Helm D (2012) *The carbon crunch: how we're getting climate change wrong – And how to fix it*, Yale University Press, London.

Hickel J, Dorninger C, Wieland H and Suwandi I (2022) 'Imperialist appropriation in the world economy: drain from the global south through unequal exchange, 1990–2015', *Global Environmental Change*, 73. https://doi.org/10.1016/j.gloenvcha.2022.102467

Hillman M, Fawcett T and Rajan SC (2007) *The suicidal planet: how to prevent global climate catastrophe*, St Martin's Press, New York.

Hobsbawm E (1977) *Revolutionaries*, Hachette, London.

Holliday J (2011) *The struggle for Syria in 2011*, Institute for the Study of War, accessed 1 June 2022. https://www.jstor.org/stable/pdf/resrep07928.1.pdf

Homer-Dixon T (2010) *The upside of down: catastrophe, creativity, and the renewal of civilization*, Island Press, Washington.

Intergovernmental Panel on Climate Change (IPCC) (2007) 'Summary for policymakers', in Solomon SD, Qin M, Manning Z, Chen M, Marquis KB, Averyt MT and Miller HL (eds) *Climate change 2007: the physical science basis – contribution of working group I to the fourth assessment report of the intergovernmental panel on climate change*, Cambridge University Press, Cambridge.

Intergovernmental Panel on Climate Change (IPCC) (2018) 'Summary for policymakers', in Masson-Delmotte VP, Zhai H-O, Pörtner D, Roberts J, Skea PR, Shukla A, Pirani W, Moufouma-Okia C, Péan R, Pidcock S, Connors JBR, Matthews Y, Chen X, Zhou MI, Gomis E, Lonnoy T, Maycock MT and Waterfield T (eds), *Global warming of 1.5°C: an IPCC Special Report on the impacts of global warming of 1.5°C above pre-industrial levels and related global greenhouse gas emission pathways, in the context of strengthening the global response to the threat of climate change, sustainable development, and efforts to eradicate poverty*, Cambridge University Press, Cambridge. https://doi.org/10.1017/9781009157940.001

Intergovernmental Panel on Climate Change (IPCC) (2021) *Climate change 2021: the physical science basis*. https://www.ipcc.ch/report/ar6/wg1/

Intergovernmental Panel on Climate Change (IPCC) (2022) *Climate change 2022: impacts, adaptation and vulnerability*. https://www.ipcc.ch/report/ar6/wg2/

Intergovernmental Panel on Climate Change (IPCC) (2023) *AR6 synthesis report: climate change 2023*. https://www.ipcc.ch/report/sixth-assessment-report-cycle/

International Peace Institute (IPI) (2016) 'Rosand: UN role in preventing violent extremism "more important than ever"', International Peace Institute, accessed 1 June 2022. https://www.ipinst.org/2016/12/preventing-violent-extremism-challenges#6

Jackson R (2018) 'Post-liberal peacebuilding and the pacifist state', *Peacebuilding*, 6(1): 1–16.

Jackson T (2009) *Prosperity without growth: economics for a finite planet*, Earthscan, London.

Jamail D (2019) *The end of ice: bearing witness and finding meaning in the path of climate disruption*, New Press, New York City.

Klein N (2014) *This changes everything: capitalism vs. the climate*, Simon & Schuster, New York.

Kolbert E (2014) *The sixth extinction: an unnatural history*, Bloomsbury Publishing, London.

Kovel J (2007) *The enemy of nature: the end of capitalism or the end of the world?* Zed Books, London.

Kunstler JH (2007) *The long emergency: surviving the end of oil, climate change, and other converging catastrophes of the twenty-first century*, Grove Press, New York.

Kurlander E (2017) *Hitler's monsters: a supernatural history of the Third Reich*, Yale University Press, New Haven.

Leahy T, Bowden V and Threadgold S (2010) 'Stumbling towards collapse: coming to terms with the climate crisis', *Environmental Politics*, 19(6): 851–868.

Lenin VI (1980) 'The collapse of the second international', in Katzer Julius (ed), *Collected works*, Progress Publishers, Moscow.

Levene M, Johnson R and Roberts P (2010) *History at the end of the world?: history, climate change and the possibility of closure*, Humanities-Ebooks.

Lilley S, McNally D, Yuen E and Davis, J (2012) *Catastrophism: The apocalyptic politics of collapse and rebirth*, PM Press, Oakland.

Linke AM and Ruether B (2021) 'Weather, wheat, and war: Security implications of climate variability for conflict in Syria', *Journal of Peace Research*, 58(1): 114–131.

Linkola P (2011) *Can life prevail?* Arktos, London.

Loadenthal M (2022) 'Feral fascists and deep green guerrillas: infrastructural attack and accelerationist terror', *Critical Studies on Terrorism*, 15(1): 169–208.

Lovelock J (2014) *A rough ride to the future*, Abrams, New York.

Lynas M (2005) *High tide: how climate crisis is engulfing our planet*, HarperCollins, London.

Lynas M (2011) *The god species: saving the planet in the age of humans*, HarperCollins, London.

MacCulloch R (2001) 'What makes a revolution?', *London School of Economics Working Paper*, 2(2001/02): 1–48.

Mach KJ, Kraan CM, Adger WN, Buhaug H, Burke M, Fearon JD, Field CB, Hendrix CS, Maystadt JF, O'Loughlin J and Roessler P (2019) 'Climate as a risk factor for armed conflict', *Nature*, 571(7764): 193–197.

Malm A (2021) *White skin, black fuel: on the danger of fossil fascism*, Verso Books, Brooklyn.

Marshall G (2015) *Don't even think about it: why our brains are wired to ignore climate change*, Bloomsbury, London.

Martell L (1994) *Ecology and society: an introduction*, University of Massachusetts Press, Amherst.

Martin P and Walters R (2013) 'Fraud risk and the visibility of carbon', *International Journal for Crime, Justice and Social Democracy*, 2(2): 27–42.

Mason P (2012) *Why it's kicking off everywhere: the new global revolutions*, Verso Books, London.

Mayerfield J (1999) *Suffering and moral responsibility*, Oxford University Press, Oxford.

McDonald M (2013) 'Discourses of climate security', *Political Geography*, 33: 42–51.

McDonald M (2018) 'Climate change and security: towards ecological security?', *International Theory*, 10(2): 153–180.

McKibben B (2007) *Deep economy: the wealth of communities and the durable future*, Holt Paperbacks, New York.

McKibben B (19 July 2012), 'Global warming's terrifying new math', *Rolling Stone*, accessed 1 June 2022. https://www.rollingstone.com/politics/politics-news/global-warmings-terrifying-new-math-188550/

McNall SG (2012) *Rapid climate change: causes, consequences, and solutions*, Routledge, London.

McPherson GR (2013) *Going dark*, PublishAmerica, Baltimore.

Meadows DH, Meadows DL, Randers J, Behrens WW (1972) *The limits to growth: a report for the Club of Rome's project on the predicament of mankind*, Universe Books, New York.

Mehta L, Huff A and Allouche J (2019) 'The new politics and geographies of scarcity', *Geoforum*, 101: 222–230.

Mering N (2021) *Awake, not woke: a Christian response to the cult of progressive ideology*, Tan Books, Charlotte.

Monbiot G (2007) *Heat: how we can stop the planet burning*, Penguin, London.

Montgomery DR (2012) *Dirt: the erosion of civilizations*, University of California Press, Oakland.

Moore S and Roberts A (2022) *The rise of ecofascism: climate change and the far right*, Polity Press, Cambridge.

Mousseau F (2019) *Evicted for carbon credits: Norway, Sweden, and Finland displace farmers for carbon trading*, The Oakland Institute, accessed 1 June 2022. https://www .oaklandinstitute.org/evicted-carbon-credits-green-resources

Murphy C (2015) 'Dignity, human security and global governance', *Journal of Human Security Studies*, 4(1): 1–12.

Nadeau R (2006) *The environmental endgame: mainstream economics, ecological disaster, and human survival*, Rutgers University Press, New Brunswick.

Nordhaus WD (1977) 'Economic growth and climate: the carbon dioxide problem', *The American Economic Review*, 67(1): 341–346.

O'Boyle G (2002) 'Theories of justification and political violence: examples from four groups', *Terrorism and Political Violence*, 14(2): 23–46.

Ophuls W (1977) *Ecology and the politics of scarcity: prologue to a political theory of the steady state*, WH Freeman, San Francisco.

Oreskes N and Conway EM (2010) *Merchants of doubt: how a handful of scientists obscured the truth on issues from tobacco smoke to global warming*, Bloomsbury Press, New York.

Oreskes N and Conway EM (2013) 'The collapse of western civilization: a view from the future', *Daedalus*, 142(1): 40–58.

O'Riordan T (1981) *Environmentalism*, Pion Ltd, London.

Orr DW (2009) *Down to the wire: confronting climate collapse*, Oxford University Press, Oxford.

Parenti C (2011) *Tropic of chaos: climate change and the new geography of violence*, Nation Books, New York.

Parrique T (2022) *Degrowth in the IPCC AR5 WGII*, accessed 1 November 2022. https:// timotheeparrique.com/degrowth-in-the-ipcc-ar6-wgii/

Pearce F (2006) *When the rivers run dry: water, the defining crisis of the twenty-first century*, Eden Project Books, London.

Peoples C and Vaughan-Williams N (2020) *Critical security studies: an introduction*, Routledge, London.

Pérez F and Esposito L (2010) 'The global addiction and human rights: insatiable consumerism, neoliberalism, and harm reduction', *Perspectives on Global Development and Technology* 9(1): 84–100.

Pörtner H-O, Roberts DC, Tignor M, Poloczanska E, Mintenbeck K, Alegría A, Craig M, Langsdorf S, Löschke S, Möller V, Okem A and Rama B (eds) (2022) *Climate change 2022: impacts, adaptation and vulnerability – Contribution of working group II to the sixth assessment report of the intergovernmental panel on climate change*, Cambridge University Press, Cambridge.

Quinn D (2000) *Beyond civilization: humanity's next great adventure*, Random House, New York.

Raskin P, Banuri T, Gallopin G, Gutman P, Hammond A, Kates R and Swart R (2002) *Great transition: the promise and lure of the times ahead*, Stockholm Environment Institute, Boston.

Richards I (2020a) *Neoliberalism and neo-jihadism: propaganda and finance in Al Qaeda and Islamic State*, Manchester University Press, Manchester.

Richards I (2020b) ''Sustainable development', counter-terrorism and the prevention of violent extremism: right-wing nationalism and neo-jihadism in context', in Blaustein J, Fitz-Gibbon K, Pino NW and White R (eds) *The Emerald handbook of crime, justice and sustainable development*, Emerald Publishing Limited, West Yorkshire.

Richards I (2023) 'Capturing the environment, security, and development nexus: intergovernmental and NGO programming for violent and hateful extremism during the climate crisis', *Conflict, Security & Development*. https://doi.org/10.1080/14678802.2023.2211019.

Richards I, Jones C and Brinn G (2022) 'Eco-fascism online: conceptualizing far-right actors' response to climate change on Stormfront', *Studies in Conflict & Terrorism*. https://doi.org/10.1080/1057610X.2022.2156036

Roelfsema M, den Elzen M, Höhne N, Hof AF, Braun N, Fekete H, Böttcher H, Brandsma R and Larkin J (2014) 'Are major economies on track to achieve their pledges for 2020?: an assessment of domestic climate and energy policies', *Energy Policy*, 67: 781–796.

Ross AR and Bevensee E (2020) 'Confronting the rise of eco-fascism means grappling with complex systems', *CARR*, accessed 1 June 2022. https://www.radicalrightanalysis.com/2020/07/07/carr-research-insight-series-confronting-the-rise-of-eco-fascism-means-grappling-with-complex-systems/

Saleam J (1999) *The other radicalism: an inquiry into contemporary Australian extreme right ideology, politics and organisation 1975–1995* [PhD thesis]. https://ses.library.usyd.edu.au/bitstream/handle/2123/807/adt-NU20020222.14582202whole.pdf;jsessionid=B8E56F208AD9A431F7820B38A025AD71?sequence=1

Shearman D and Smith JW (2007) *The climate change challenge and the failure of democracy*, Praeger, Westport.

Shiva V (2013) *Making peace with the earth*, Pluto Press, London.

Simms A (2013) *Cancel the apocalypse: the new path to prosperity*, Hachette, London.

Smith E, Persian J and Fox VJ (2023) 'Introduction: fascism and anti-fascism in Australian history', in Smith E, Persian J and Fox VJ (eds) *Histories of fascism and anti-fascism in Australia*, Routledge, London.

Smith R (2010) 'Beyond growth or beyond capitalism', *Real World Economics Review*, 53(2): 28–36.

Solnit R (2010) *A paradise built in hell: the extraordinary communities that arise in disaster*, Viking Press, New York.

Speth JG (2008) *The bridge at the edge of the world: capitalism, the environment, and crossing from crisis to sustainability*, Yale University Press, New Haven.

Spratt D and Sutton P (2008) *Climate 'code red': the case for a sustainability emergency*, Scribe Publishers, Melbourne.

Steffen W and Griggs D (2013) 'Compounding crises: climate change in a complex world', in P Christoff (ed) *Four degrees of global warming: Australia in a hot world*, Routledge, London.

Swyngedouw E (2010) 'Apocalypse forever?: post-political populism and the spectre of climate change', *Theory, Culture & Society*, 27(2–3): 213–232.

Szenes E (2021) 'Weaponizing the climate crisis: the nexus of climate change and violent extremism', *Norwich university news edition: voices on peace and war*, accessed 1 June 2022. https://www.norwich.edu/news/voices-from-the-hill/peace-and-war/3548 -weaponizing-the-climate-crisis-the-nexus-of-climate-change-and-violent-extremism

Tainter J (1990) *The collapse of complex societies*, Cambridge University Press, Cambridge.

Taylor B (2008) 'The tributaries of radical environmentalism', *Journal for the Study of Radicalism*, 2(1): 27–61.

Thomas C and Gosink E (2021) 'At the intersection of eco-crises, eco-anxiety, and political turbulence: a primer on twenty-first century ecofascism', *Perspectives on Global Development and Technology*, 20(1–2): 30–54.

Thomas KA and Warner BP (2019) 'Weaponizing vulnerability to climate change', *Global Environmental Change* 57: https://doi.org/10.1016/j.gloenvcha.2019.101928

Tilly C (1978) *From mobilization to revolution*, Addison-Wesley, Boston.

Trainer T (2011) 'The radical implications of zero growth economy', *Real-World Economics Review*, 57(1): 71–82.

Trainer T (2012) 'De-growth: do you realise what it means?' *Futures*, 44(6): 590–599.

United Nations (UN) (2021) *Our common agenda: report of the Secretary-General*. https://www.un.org/en/content/common-agenda-report/assets/pdf/Common_Agenda_Report _English.pdf

United Nations Development Programme (UNDP) (2020) *UNDP policy brief: the climate security nexus and the prevention of violent extremism – Working at the intersection of major development challenges*. https://www.undp.org/sites/g/files/zskgke326/files/ publications/UNDP-Climate-Security-Nexus-and-Prevention-of-violent-extremism.pdf

United Nations Office for Disaster Risk Reduction (UNDRR) (2022) *Global assessment report on disaster risk reduction 2022: our world at risk – Transforming governance for a resilient future*, Geneva. https://www.undrr.org/publication/global-assessment-report -disaster-risk-reduction-2022

Urry J (2011) *Climate change and society*, Palgrave Macmillan, London.

Von Daniken E (2018) *Chariots of the gods*, Penguin, London.

Walker J (2010) *Disaster utopianism: looking for paradise in catastrophic places*, Reason, https://reason.com/2010/04/20/disaster-utopianism/ .

Wallace-Wells D (2020) *The uninhabitable earth: life after warming*, Crown, New York.

Walt SM (16 January 2011) 'Why the Tunisian revolution won't spread', *Foreign Policy*, accessed 1 June 2022. https://foreignpolicy.com/2011/01/16/why-the-tunisian-revolution -wont-spread/

Warner J and Boas I (2019) 'Securitization of climate change: how invoking global dangers for instrumental ends can backfire', *Environment and Planning C: Politics and Space*, 37(8): 1471–1488.

Weis VV and White R (2020) 'A Marxist perspective on the 2030 Agenda for Sustainable Development', in Blaustein J, Fitz-Gibbon K, Pino NW and White R (eds) *The Emerald handbook of crime, justice and sustainable development*, Emerald Publishing Limited, West Yorkshire.

Weisman A (2013) *Countdown: our last, best hope for a future on earth?* Hachette, London.

Winston A (2014) *The big pivot: radically practical strategies for a hotter, scarcer, and more open world*, Harvard Business Review Press, Boston.

Wong C, Gold S, Rizk S and Flynn C (2020) 'Re-envisioning climate action to sustain peace and human security', Prevention Web, accessed 1 June 2022. https://www.preventionweb .net/news/re-envisioning-climate-action-sustain-peace-and-human-security

Woodbridge RM (2004) *The next world war: tribes, cities, nations and ecological decline*, University of Toronto Press, Toronto.

World Meteorological Organization (WMO) (2022) *Provisional state of the global climate*. https://library.wmo.int/doc_num.php?explnum_id=11359

Žižek S and Park Y (2013) *Demanding the impossible*, Polity Press, Cambridge.

Zovanyi G (2013) *The no-growth imperative: creating sustainable communities under ecological limits to growth*, Routledge, New York.

6 Far-right (anti-)environmentalism in the post-truth era

Global networks and future directions

Introduction

This book has examined a circumstance in which authoritarian and ethnonationalist political perspectives on climate change represent the broader inculcation of far-right ideas into mainstream political settings (Dyett and Thomas 2019; Thomas and Gosink 2021). It has explored contemporary far-right responses to climate change and wider environmental politics expressed by Australia-based far-right actors, both historically and in the present. The various right-wing traditions of climate change politics in question were shown to include accelerationism, alongside resignatory and denialist responses to climate change mitigation measures and to the devastating international impacts of global heating (Richards 2023). The analysis demonstrated that although the various traditions of climate response are often considered discrete, when examined against the backdrop of far-right actors' wider uses of environmentalist politics, they are in certain respects historically continuous, sharing culturally and socioeconomically exclusivist elements.

Several cases of (pseudo-)environmentalist messaging on the part of far-right actors in Australia were shown to feature notions of land and place connection in association with conspiratorial narratives of (white) demographic 'replacement' (Camus 2012). As discussed in some detail in Chapter 2, this was perhaps unsurprising, given the country's origins as the European colonial settlement of 'White Australia' (Lindqvist 2007). Due to this history, Australia has been recognised by leaders of white supremacist movements internationally for its strong overt and covert cultural and political resonances with white supremacist and proto-fascist ideology. As Oswald Mosley, the leader of the British Union of Fascists, stated in 1933: 'I always thought it remarkable that Australia, without studying the Fascist political philosophy and methods, so spontaneously developed a form of fascism peculiarly suited to the needs of the British Empire' (Tully 2015).

Replacement narratives championed by the Australian far right were partly inspired by the writings of European New Right thinker Renaud Camus. In certain respects, they could be said to echo Camus's suggestion that global(ist) elites, usually implicitly codified as Jewish, are deliberately seeking the demographic replacement of white people by people of colour through the processes of migration, procreation between people of different cultural backgrounds and 'outbreeding' of white people generally (Richards et al. 2021). Other elements of the replacement

DOI: 10.4324/9781003325437-6

theory also important in the rhetoric of the Australian far right include the conspiratorial belief that this process of replacement features concerted efforts to destabilise the societal dominance of the traditional family unit through the coordinated promotion of egalitarian progressivist politics, often by metropolitan elites (Roose 2020; Abbas et al. 2022).

On the end of the far-right continuum, but tending towards the radical rather than the extreme right (Minkenberg 2017; Mudde 2019), fears of a loss of white masculinist privilege associated with replacement theory justify the radical instauration of nationalistic white power. This has been expressed in the Australian context through cultural xenophobia, interpersonal and structural violence enacted against non-white people, and the societal proliferation of radical and exclusionary anti-immigration sentiments. On the other, more immediately violent and 'extremist' end of the spectrum (Carter 2013), these belief systems are deployed to justify homicidal and would-be genocidal responses to the Great Replacement; this conspiracy theory being the title of the so-called manifesto of Brenton Tarrant, the Australian neo-fascist who murdered 51 Muslims and injured at least 40 more in a mass shooting in Christchurch in 2019 (Macklin 2019).

The normalisation of replacement narratives and status anxiety on the part of organised white nationalists is not limited to Australia, these tropes being increasingly prevalent in the statements of politicians and mainstream news and social media commentators (Obaidi et al. 2022). In 2021, for instance, US Fox News host Tucker Carlson alleged a conspiracy on the part of the DNC to promote the demographic substitution of white Americans through international migration for the purpose of generating pro-globalisation DNC voting blocs (Ramirez 2023). The Great Replacement was also the title and subject of a documentary produced by the white nationalist social media influencer Lauren Southern (2017), who now enjoys a semi-regular talking spot on Sky News Australia. In a clear example of the forces that lead to far-right mainstreaming, as of August 2021, Sky News Regional has been broadcast free to air in non-metropolitan parts of NSW, Victoria, Queensland and South Australia. This occurred in the same year as a Senate Inquiry into Media Diversity highlighted YouTube's ban on Sky News Australia for spreading misinformation and contravening its policies on hate speech, including, significantly, the Sky News interview feature with high-profile neo-Nazis in Australia and spread of white supremacist views in its night-time programming, for which it earned the pejorative local moniker 'Sky News after Dark' (Dixon 2021).

In exploring the domestic and international political roots of Australian far-right actors' politically violent attitudes, this research has sought to foreground the influence of contemporary narratives of supremacy and oppression and historical contexts of material deprivation. This critical approach to understanding oppressive politics should be understood as problematising conventional notions of 'extremism', where societal stability is pursued as a state-centric goal, denying the necessity of radical social and economic change to mitigate the global impacts of social crises, such as rising neo-fascism, and counterfactual media coverage exacerbating inequality and extreme climate devastation (Martini et al. 2020;

Walker 2021). In exploring these dynamics and the events they speak to, this book has sought to examine the factors that compel radical responses to global heating, thereby illuminating the dynamic relationships between 'extreme' and 'mainstream' political forces and seeking redress for ahistorical interpretations of the far right and its coverage in the media.

As a part of its analysis, the book has examined the ways in which contemporary securitising and development-focused counter-terrorism initiatives employed by states and multilateral institutions are often framed by the ESD nexus. Through a critical lens of inquiry, this discussion examined how securitising responses to terrorism and climate disasters have often been predicated on racism, Othering or a lack of empirical information, in turn yielding anti-humanitarian impacts and shaping future patterns of political violence (Stampnitzky 2013; Richards 2020a). Institutional–governmental responses were then explored alongside alternative approaches, including grassroots and radical responses to climate change. An examination of New Catastrophism considered its eco-socialist, eco-anarchist and progressivist ecological-environmental tendencies, including those that provide genuine alternatives to green authoritarian perspectives, as opposed to those that may be vulnerable to far-right co-optation.

The next part of this chapter reflects on the insights provided by the empirical analysis of Australian news and social media signifying far-right environmental politics. A final section then synthesises the insights set out thus far and extends upon Chapter 5's discussion of alternate possibilities for a future transformative agenda.

Historical and international connections

While this book has examined the ways in which counter-terrorism policies paired with developmentalism can produce injurious and criminogenic impacts, another key supposition guiding our investigation was that the propagandised impacts of far-right activity and its coverage in news media are co-produced, often leading to inadvertent relationships of reciprocity and interdependence between media and far-right actors (Howie and Campbell 2017; Sparrow 2021). This approach does not assume any moral equivalence or false comparison between different actors, but rather seeks to understand from a peace and conflict studies perspective how dynamics of violence are dialectically perpetuated through their mediatised expression (see Jackson 2015; Bloom 2016). Examining the relationships between purportedly opposing 'mainstream' and 'extreme' entities in a holistic fashion represents a critical alternative to isolated, atomised interpretations of politically violent phenomena (Jackson 2015). By revealing the pathways to right-wing radicalisation, our approach could be said to empower those seeking to prevent this form of political violence, by providing them with an avenue through which to influence the course of events within situations of political violence and devastation (Lindahl 2018).

When considering how intersecting social, economic and environmental crises such as climate change are routinely exploited by the far right, it is also important to recognise how far-right efforts at propagandising and recruitment have always been syncretic (Eatwell 1996; Griffin 2013). With an emphasis on far-right efforts at entryism and recruitment, including targeting the political left, Chapter 2 explored some of the key historical engagements with environmental politics by radical, far-right and extreme-right social forces in Australia. Since Federation and the 1901 Immigration Restriction Act, the use of syncretism by the Australian far right has involved anti-imperialist, nativist appeals to the exclusive rights of white workers (Burgmann 1984). Such syncretism has also been evident in the early AFM and the 'Jindyworobak' literary club in the interwar period superficially appropriating Indigenous imagery and concepts, and blending them with hybridised Aryan and Odinist ideas of 'race and place', alongside quasi-Eurasianist appeals of land and place connection (Stephensen 1935; Bird 2014). The tenets of deep ecology influential on US 'green' and 'brown' tribal movements since the 1980s were then shown to be important for self-described NAs in the UK in the 1990s and then Australia in the early 2000s, although these actors rejected anarchism's political–philosophical morale of egalitarianism, instead advocating hierarchical, localised ethnic communities governed according to substate nationalist and authoritarian–masculinist principles (Macklin 2005; Sunshine 2008).

Although the Australian far right did in some ways uniquely develop their ideology within the political–ideological framework of the White Australia policy, they were also inspired by and adapted other common ideological tendencies of far-right white nationalism (Smith 2020). The white nationalist tropes important to early competing republican and imperialist political movements in Australia, for example, have been described as reminiscent of a twentieth-century 'red-brown' blending of socialistic and national–fascistic ideas in other international contexts. These included some anti-capitalist Nazis such as the National Bolshevik party and Strasserism, the Russian National Bolshevik mode of political organisation during the Cold War, and Aleksandr Dugin's more recent Fourth Political Theory (Sunshine 2008), which was combined in some disinformation media networks with the anti-capitalist and anti-US imperialist tendencies of the post-1968 European New Right (Bar-On 2016; Bevensee 2020).

White nationalist political movements within Australia and internationally during the Cold War period, then, incorporated the ideological tenets of global anti-communism, as well as ruralist notions of anti-urbanism. Today many of these same ideas connect with anti-globalisation narratives of the white working class being 'left behind'. Left-behind belief systems propagated by right-wing actors tend to argue that the offshoring of industrial manufacturing and other mechanisms of cultural and economic globalisation have contributed to the 'ghettoisation' of working-class communities in the 'West', either implicitly or explicitly signifying a decline in the living standards of white people (Antonucci et al. 2017; Mondon and Winter 2020). In the differentiated fashion of the ENR and quasi-Eurasianist actors (Horvath 2013), these notions of economic disenfranchisement are often also accompanied by stated opposition to US and UK imperialism, as epitomised in the Carthaginian or Atlanticist sea-based economic world order (Antón-Mellón 2013). Despite the various anti-capitalist tenets of these social movements, they also tend to hold opposition to 'socialism' in its non-nationalist internationalist variants. The propagandised application of historical, internationalised ideas in Australian far-right actors' environmentalist narratives also sometimes leverages sustainability narratives, combined with genuine concern with climate change through more and less overt appeals to 'green authoritarian' solutions (Ross and Bevensee 2020).

As highlighted at several points in the discussion, the synthesis of disparate political positions and ideas in the service of a more immediate white supremacist agenda has significant historical precedent. The far-right Cultural Marxism conspiracy theory prevalent across far-right Australian and international media today derives from historical Judeo-Bolshevism and Cultural Bolshevism theories, which themselves developed from earlier nationalistic forms of antisemitic hatred in the century leading up to WWII (Busbridge, Moffitt and Thorburn 2020). While the Judeo-Bolshevism theory holds that Jews led the 1917 Russian Revolution, and the Cultural Marxism trope suggests that the Frankfurt School of Critical Theory advocated cultural infiltration and subversion rather than economic action to foment revolution in the contemporary context, collectively these theories cast Jewish people as 'borderless enemies' who promote gender and sexual perversion, and seek to undermine traditional Christian or national values and economies (Hanebrink 2018). While the influence of Cultural Marxism remains prevalent, so, too, are other, more explicitly neo-Nazi antisemitic conspiracies related to the Zionist Occupation Government or 'ZOG' narrative. Originally supposedly conceptualised by the US neo-Nazi Eric Thompson in a 1976 magazine article 'Welcome to ZOG-World', and popularised by leader of the US National Alliance, William Luther Pierce, in 1978, this narrative alleges that Western neo-Marxists and their Jewish apparatchiks seek the dissolution of Western civilisation and its traditional institutions by exerting financial control over national governments (Shekhovtsov 2013).

Australian far-right propaganda in cross-national online media was demonstrated as holding prominent engagement with the ideological descendants of ZOG, such as Renaud Camus' the Great Replacement, combined with agitation against liberal capitalism, alongside socialism, communism and Cultural Marxism. Particularly across Gab, Stormfront and Australian far-right alt-news media, climate change governance was interpolated into a metanarrative – often alongside policies for COVID-19 pandemic management – of a corporate global(ist) elite's attempting tyrannical control and manipulation of national populations (Richards 2022). The historical foundations of these ideas in Australia include the actions of the longest-running far-right organisation in the country's political history, the ALR, whose virulent antisemitism accompanying its semi-successful entry into the Liberal and Country (now National) Party in Australia was revealed by Ken Gott's 1965 *Voice of Hate*, among other studies (Moore 2005). White supremacy and antisemitic signposting of 'international capitalists' are also evident in the agitations of various Labor MPs, from the first ALP Premier of NSW JT McGowan's statement that 'While Great Britain is behind us, and while her naval power is supreme, Australia will be what Australians want it – white, pure and industrially good' (Nairn 1969, 10); to ALP MP Frank Anstey's texts, *The Money Power* (1921) and *The Kingdom of Shylock* (1917). In light of the reflexive populationist-eugenicist focus of contemporary green nationalists and ecofascists, the longstanding far-right allegations of tyrannical, 'biopolitical' population control enacted by globalists might be interpreted as somewhat ironic.

Fear about 'racial miscegenation' and the 'deracination' of white communities has a unique Australian flavour, being often expressed through historically resonant fearmongering about threats from 'the East' (Walker and Sobocinska 2012). A well-known example is the contemporary uses of the Eureka flag by neo-Nazi actors that memorialise nineteenth-century European goldminers' labourist agitation against their imperialist masters, and their violent racial targeting of Chinese goldminers (Platoff and Knowlton 2015). Yellow Peril narratives were also important for Australian political institutions throughout their early history (see Chapter 2, p. 51, Chapter 3, p. 89), as demonstrated by former ALP Member of Parliament for Kalgoorlie, Graeme Campbell, who went on to found the far-right Australia First Party in June 1996 and joined Pauline Hanson's ONP in 2001. In October 1996, Campbell presented a conference paper in the Australian House of Representatives penned by Denis McCormack, titled 'The Grand Plan, Asianisation of Australia – Race, Place and Power' (The Parliament of the Commonwealth of Australia 1996).

In their crudest forms, white nationalist anxieties expressed in Australian parliamentary political rhetoric have also at times combined with antisemitic or seemingly quasi-Nazi attitudes, remediated for the local context. In the 1990s and 2000s, the biography of far-right ONP leader, *Pauline Hanson: the Truth*, referred to 'the Aboriginal Question' (Hanson and Merritt 1997), recalling the German Nazi regime and neo-Nazis propagandising about 'the Jewish Question' (Gordon 1984). More recently, such remarks have grown even more extreme and explicit – another Australian senator, Fraser Anning,[1] declared in his maiden parliamentary speech in 2018 that Australia required a 'final solution' to the 'problem' of non-white immigration (Conifer 2018).

'Post-truth' and new media

Propagandising communications and statements of intent by different far-right actors must be contextualised not only in relation to the national historical context in which they were produced, but also in terms of their constructivist messaging. In 2022, the authors of this book examined discussions of climate change on Stormfront, focusing on the ways that platform users leveraged prevalent political discourse about the concept of 'ecofascism' to construct a politically convincing message (Richards et al. 2022). That investigation highlighted Stormfront users' debates over what ecofascism is, and what function it might serve in propagandising for recruitment to white supremacist and genocidal causes.

Drawing on notable historical cases and examples, this book has also sought to show how counterfactual narratives promulgated in response to climate change by differentiated Australian far-right actors are encouraged not only by certain political histories but also by several intersecting structural–cultural features of the media through which they are expressed. These include (a) the highly concentrated ownership of Australian news media, and the intertwined commercial arrangements and political–economic interests of news and social media producers and owners (see Richards et al. 2020); (b) the monetisation of viral media online, which exploits psychologically addictive material for commercial purposes and encourages a lack of nuanced debate on critical social issues, partly by encouraging clickbait, sensationalism and transitory patterns of engagement (Charkawi et al. 2021); and (c) neoliberal intellectual environments that contribute to the cultural atomisation of vulnerable internet users, while simultaneously attaching instrumental and commercial values to the notion of 'truth' itself, including scientific truth on anthropogenic climate change (Biesecker 2018; Giroux 2020).

Considering the influence of contemporary mediatised environments on Australian far-right actors' uses of environmentalist politics, this book has also made the case for recognising the ideological provenance of contemporary anti-knowledge politics about climate science, combined with pseudoscientific ideas on 'race'. Although far-right political philosophy was not a primary focus of this book, contiguity between climate change denialism, resignation and accelerationism on the part of contemporary far-right actors can partly be explained by the influence of proto-fascist ideology on contemporary 'post-truth' environments. The term 'post-truth' was named the 2016 Word of the Year by *Oxford Dictionaries*, and this period is marked by the dominance of ideological beliefs among political and economic elites. This dominance often involves the rejection and dismissal of well-founded knowledge and evidence, generally with the aim of reinforcing social hierarchies or advancing political objectives (Biesecker 2018; McIntyre 2018).

Social psychological explanations of post-truth are more common than structural accounts (Haidt 2012; Stanovich et al. 2016), and the political–philosophical treatment of 'truth' by far-right actors in what are often described as post-truth settings is often overlooked. To understand far-right and proto-fascist movements' lack of interest in evidence or proof, or lack of concern about their own hypocrisy, it is useful to examine their embrace of Social Darwinist 'survival of the fittest' ideologies. These ideologies often integrate Platonian ideas that 'might equals right' (Gentile 2003; Paxton 2007; Griffin 2013), combined with Nietzschean notions of the Will to Power, where the veneration of physical prowess and spirit of domination is seen to restore cultural destiny or greatness, 'beyond Good and Evil' (Bataille 2015; Wolin 2019).

Indeed, although early Australian myths might not be described as fascist, some of these ideas resonate in the Australian cultural context and extreme-right actors exploit these cultural narratives as such, which is particularly reinforced by the prevalence of a national mythology of a 'rascal nation' rejecting cultural difference and valuing physicality above intellectual pursuits (Ward 1978). In this context, the vigour and strength of youthful white men representing the Australian population developed in part through the notion of a Nietzschean 'spiritual essence' (see Lukács 1952), exemplified in early rural workers' symbolic taming of wild Australian lands (Curthoys 1999; Veracini 2007). With a perhaps comparable, regionalist opposition to urbanism, through Fascist Italy in the 1920s and the lead-up to WWII in Germany during the Weimar Period, anti-intellectual values rejecting an academic elite as superior to 'the people' developed as an extension of Romantic and reactionary responses to both cultural progressivism in respect of socialist institutions and the positivistic influences of the Enlightenment (Gregor 1979; Glaser 2019). These developments also had early pre-fascist roots in a cultural revolt against the positivism that predominated in many European countries in the late nineteenth century (Sternhell et al. 1994).

Chapter 4 also highlighted how it is important in the contemporary Australian context to recognise that post-truth and the related concept of 'fake news' are contested terms, since both lack stable meaning, are usually associated with injurious politics and problematically assume the existence of inclusive or representative 'real news' (Habgood-Coote 2019; Mondon and Winter 2020). Despite the transformative impacts of new media environments, news media coverage of social oppression and controversy has long been conducive to aims of political domination, undermining the in-principle goals of public information campaigns, despite the emancipatory or democratic intentions of many journalists (McChesney 2016). Positivist interpretations of 'truth' may also reflect the limitations that arise from structural bias and political–economic imperatives in news media industries, afflicting the (democratically vital) role of news communication in various broadcast arenas (Schiller 1973) – and the Australian media sector experiences obstacles to transparency in a particularly pronounced way. Despite these issues, there are of course still elements within contemporary media spaces that undeniably do disseminate and circulate deliberately misleading and often outright false claims for various reasons, including for political and ideological gain.

This book's analysis treats mainstream news and social media as vital agents of propaganda, and emphasises the adverse consequences associated with both greater and lesser coverage of nascent politics on the far and extreme right. While some patterns of coverage can either amplify or obscure the foundations of far-right messaging (Brown et al. 2021), other journalistic tendencies in Australia, as elsewhere, have supposed it harmful to cover the ideology or political strategies of far-right and neo-fascist actors, or have avoided doing so because of cultural taboos, or a lack of perceived threat (Paxton 2007; Rydgren 2018). As Sparrow (2021) highlights, drawing on cases and examples of media reportage on the far right in Australia, this can, in turn, counterproductively contribute to a circumstance in which neo-fascist and far-right activists can at once promote their ideas, organisations and events while downplaying the most pernicious elements of their belief systems.

This book has responded to burgeoning, sensationalised coverage of both far-right social movements and climate change (Wodak 2015; Mann 2021) by suggesting that to prevent exclusivist and supremacist politics from becoming more prevalent, it is necessary to recognise their wider cultural context. This includes acknowledgement of the political spectrum to which they belong, which features a degree of alignment between disparate extreme and more mainstream political positions, as well as elements of contestation and co-optation between far-right and other radical typologies of climate response. Mainstream–extreme alignment is apparent, for example, in civilisationist attitudes to population control connected with climate change, which are not only held by adherents or perpetrators of political violence (Moore 2015). Contemporary New Right and identitarian advocacy of 'remigration' is also becoming more prevalent on the far right, signifying policies for the forced repatriation of regular and irregular migrants from Europe and European settler countries to their alleged ancestral homelands (Mudde 2017; Zúquete 2018).

Other increasingly mainstream exclusionary attitudes in response to the effects of climate change are anti-immigrant 'fortress' policies, referring to a set of political measures aimed at limiting migration, strengthening border controls and restricting the rights of migrants already present in a country (Kaldor and Sassen 2020). These policies are often characterised by their emphasis on security and the use of force to prevent the entry of irregular or unauthorised migrants (Jacobson 2019). 'Fortress Europe' or 'Fortress Australia' mentalities and policies encouraged by some national media, and increasingly by national governments, respond in part to conditions of impending climate devastation, in relation to which refugees are conceptualised as security risks (Mountz 2010; Baldwin et al. 2014).

Restrictive approaches to borders and migration have special meaning for the island state of Australia, given its geopolitical history, and successive Australian governments' recent historical legal and discursive criminalisation of irregular migrants (Crock et al. 2012; Gerard and Weber 2019; McAdam 2013). Indeed, the brutality of Australia's approaches to borders and migration was made apparent in 2015, a report by UN Special Rapporteur on Torture, Juan Mendez, declared in relation to refugee processing on Manus Island that 'Australia is systematically violating the international Convention Against Torture by detaining children in immigration detention', and 'The Migration and Maritime Powers Legislation Amendment ... violates the Convention Against Torture because it allows for the arbitrary detention and refugee determination at sea, without access to lawyers' (Doherty and Hurst 2015). In 2017, the UN High Commissioner for Refugees, Filippo Grandi, expressed similar concern about the prolonged detention of refugees and asylum seekers in facilities on Nauru and Manus Island, citing reports of abuse and neglect (UNHCR 2021).

The international impact of Australia's restrictive approaches to border control was also apparent following the 2013 implementation of the then LNP Coalition government's 'Operation Sovereign Borders' – which was endorsed through campaign messaging by successive governments, including the 2022 ALP Albanese administration, and remains in force at the time of writing (Perera 2009; Boochani 2018). This programme followed a militarisation of Australia's border policies in the aftermath of the Al-Qaeda attacks on the Washington Pentagon and New York World Trade Center in 2001 (McCulloch 2004). In that case, the Liberal party under John Howard campaigned successfully in the 2001 Australian federal election on the basis of fallacious news-media reporting about asylum seekers throwing their children overboard from the Tampa vessel moored off of Christmas Island (Martin 2015). As Australian troops boarded the vessel, the Coalition signposted perceived security risks associated with refugees from the Middle East (McKay et al. 2011).

Operation Sovereign Borders entailed the further militarisation of Australian maritime borders, the establishment of offshore detention facilities, and a concerted messaging campaign to deter those who would transport asylum seekers to Australia, with a particular emphasis on those who arrived by boat.[2] A YouTube Border Force video advertisement featured Australian General Angus Campbell stating in a direct address to potential asylum seekers: 'No way: You will not make Australia home' (ABF TV n.d.; Richards 2022, 12). This video and the policy messaging were directly praised and appropriated by influential European far-right political forces. The neo-fascist youth movement Generation Identity, for example, displayed banners stating, 'No Way ... You Will Not Make Europa Home' in its 'Defend Europe' missions (@GID_UKIRE n.d.; RT Producers n.d.). In 2017, the movement's first mission involved interfering with NGO vessels seeking to rescue migrants drowning in the Mediterranean Sea, and its second mission in 2018 planned to intercept migrants at the Swiss Alps border. Geert Wilders of the far-right Dutch Party for Freedom included within his party's anti-immigration policy campaign a video bearing almost identical nautical imagery to the Australian video, with Wilders stating the phrase 'No way: you will not make the Netherlands Home' (PVVpers 2015; see Richards 2022, 12).

The supremacist and exclusionary logic underwriting these contemporary expressions of anti-immigrant xenophobia is apparent in the colonial attitudes deployed to justify the disproportionate suffering of those in the Global South caused by the impacts of global heating, including environmental degradation and 'slow violence' (Escobar 2011; Nixon 2011). Arguments in support of fortress policies are also morally spurious, given that Global North and wealthy states developed historically on the back of the exploitation of human labour and resources from LDCs, while those countries now bear the disproportionate deleterious impacts of climate change (Blaustein et al. 2020; Richards 2020b). Moreover, as this book has sought to highlight, while eugenicist and civilisationist population control attitudes are increasingly recognised as characteristic and inspirational of 'ecofascism', often overlooked is the influence of environmental-development tropes that employ civilisational rhetoric, which are also embraced by self-described ecofascists (see Chapter 3, p. 74). The severity of these connections is starkly apparent in the white supremacist tracts of mass attackers, including Brenton Tarrant's so-called manifesto. That text called for an end to what Tarrant described as: 'This stripping of wealth and prosperity in order to feed and develop our cultural competitors is an act of civilization [sic.] terrorism resulting in the reduction in development and living conditions of our own people for the benefit of those that hate us' (2019, 42).

Population-control narratives that erroneously assign blame for climate change to the Global South also contribute to misinformation about its anthropogenic causes, and to the logic underpinning (ethno-)nationalist responses to global heating (Angus 2016). Reports produced at intergovernmental agencies often refer to net population rather than per-capita resource use as a barometer of relative climate impacts, and can therefore be put toward populationist arguments about the causes of climate change (Hickel et al. 2022). Other texts on and approaches to climate change mitigation that also find purchase on the far and extreme right, while not composed within ethnonationalist or authoritarian frameworks, contain elements that are vulnerable to far-right co-optation. As Chapter 2 explored, these include biologist Paul Ehrlich's *The Population Bomb* (Ehrlich 1968), which he wrote after visiting Delhi and complaining of 'overpopulation', despite the fact that other major city centres in the West, such as Paris, were much more densely populated at the time (Mann 2018; Yakushko and De Francisco 2022). Even the progressivist think-tank Club of Rome's 1972 report, 'The limits to growth', was vulnerable to authoritarian applications for its theorisation of a supposed imminent threat to the scarcity of natural resources (Mehta et al. 2019; see, e.g. Ophuls 1977; Heilbroner 1974). Although Club of Rome scholarship has been used to international progressivist and anti-capitalist ends, the scarcity narratives promoted in this text and elsewhere also informed intergovernmental security narratives that emphasise interstate competition for resources, the realpolitik fear of supposed 'water wars' (McDonald 2018) and the necessity of containing 'risky' populations in the Global South, as was explored in some detail in Chapter 5 (Warner and Boas 2019).

However, rather than emphasising notions of ecological security or alternative approaches to global economic systems research (McDonald 2013), the empirical and secondary analysis presented across the chapters in this book demonstrated how the wider environmental security and environmental peacebuilding literature can be vulnerable to far-right co-optation (see Chapter 5). This is due to its focus on precarity and scarcity driven by resource consumption, despoliation of the natural environment, and an alleged lack of institutional governance in LDCs to meet the needs of local populations (Busby 2018; Richards 2023). These intersecting sustainability narratives derive from recent approaches to the ESD nexus, but also from currents of thought prevalent in different national contexts and during different historical periods. Such thinking includes that of early figures in the US eugenics movement who had significant influence on public policy, such as Charles Davenport, Madison Grant and Harry Laughlin. Also influential was the early Western development of environmentalist thought, which often entailed racist ideologies and beliefs in the superiority of certain racial or ethnic groups (Guha 2014); the early scientific field of ecology, in which Frederic Clements advocated his belief in the concept of ecological succession, where ecosystems are seen to develop in linear fashion towards a 'climax community' or 'ideal state' (Merchant 1980); and the 1960s emergence of global population reduction arguments.

In this context it is relevant to note that the concept of 'sustainability' did not always relate to the instrumental ambitions set out in the UN's 2030 Sustainable Development Agenda, connoting sustainable *capitalism* (Blaustein et al. 2020). Twentieth-century strains of radical and modern environmentalism conceptualised sustainability in ecological terms, often drawing significantly from various forms of anarchism. Expanding further on the analysis in Chapter 5, we focus on different tendencies within radical environmentalism that could be said to share points of interconnection with far-right, green authoritarian responses, as well as others that could provide inspiration for an alternate climate response. Elaborating on several post-capitalist conceptions of what sustainability might mean, the following section explores futures imagined in different strains of green anarchism. These responses challenge capitalism and developmentalist colonialism, as the structural origins of the current crisis situation, which is marked by the exchange between (eco)fascism and global heating itself.

Green anarchism and post-capitalist imaginations

While Chapter 5 outlined the concepts of the ESD nexus and the different tenets of New Catastrophism, these approaches were argued to hold the potential to either respond to, provide context for, or even support authoritarian and ethnonationalist tendencies in far-right climate change politics. Both the approaches to the ESD nexus and the agitation of New Catastrophism were based on fears of an unsustainable future, with respect to resourcing, effective governance and the avoidance of undesirable social-ecological outcomes. A historical lens is useful once more for understanding the evolution of differentiated grassroots and institutional responses to global heating and how these responses have informed the far-right (often pseudo- and quasi-)environmentalist ideologies that are the focus of this book. In particular, modern environmentalism and contemporary understandings of 'sustainability' might in this respect be properly contextualised in terms of their derivation and deviation from radical environmentalism, particularly from various tendencies within green anarchism. While in recent years debates on whether capitalism is conducive to ecological sustainability have become remarkably mainstream (see IPCC 2022; IPCC 2023), many of the persistent elements in these debates have their origins in anarchist political theory.

One of the early influences on contemporary environmentalist politics that explore the possibilities of a post-capitalist world was Murray Bookchin's formative work, which firmly established the importance of ecological issues in anarchism and the relevance of the anarchist critique of authority to environmental issues. As one of the first works on modern environmentalism, Bookchin's *Our Synthetic Environment* (Bookchin 1962) critiqued the massive post-war use of industrial chemicals across society.[3] Continuing his development of radical political ideas and environmentalism in *Post-scarcity Anarchism* (Bookchin 1971), this text argued that capitalist society is incompatible with ecological survival (what is now called 'sustainability') but that a decentralised anarchist social organisation that harnesses modern technology for communal ends in harmony with ecological concerns would not need to lead to a frugal, ascetic existence, but could be one of vitality and abundance. The text was published one year before the Club of Rome's *Limits to Growth* (Meadows et al. 1972), which agitated for increased recognition of the implications of economic growth for competition over natural resources.

Bookchin developed this position into the eco-anarchist philosophy of 'social ecology' in the major works *Toward an Ecological Society* (Bookchin 1980), *The Ecology of Freedom* (Bookchin 1982) and *The Philosophy of Social Ecology* (Bookchin 1990), in which he extended the critique of authority to hierarchical relationships with non-human nature. Through continual development of this philosophy and ongoing bitter polemic debate with other anarchists, which led to him feeling disconnected from late-twentieth-century anarchism, Bookchin eventually disavowed the label anarchism, preferring to refer to his anarchistic revolutionary vision as 'communalism' or even 'municipal libertarianism' in later works such as *Social Ecology and Communalism* (2007).

Unlike the often-simplistic anti-technology (or more accurately anti-*modern*-technology) perspective of anarcho-naturism, post-scarcity anarchism is a pro-technology vision, and it has been called upon to these ends in various post-capitalist imaginations of a response to global heating (Biehl 2015; Morris 2015; Nelson 2018). Particularly influential is Bookchin's view that the best hope for an ecologically sustainable human society lies in the harnessing of modern technology (including, prophetically, information technology) to achieve the best possible quality of life and harmony with ecological systems, rather than for profit. This vision emphasised the need to consider social issues alongside environmental ones, which contrasted with the then popular ecological framework of 'deep ecology', some proponents of which had developed a misanthropic perspective. Janet Biehl and Peter Staudenmaier, two pre-eminent researchers of ecofascism, in their influential text *Ecofascism: Lessons from the German Experience* (1995) and more recently in Staudenmaier's *Ecology Contested* (2021), have endorsed Bookchin's vision, highlighting in their view the potential vulnerability of deep ecological perspectives to far-right entryism. This is not to suggest that deep ecology is intrinsically and inescapably fascistic, however. As Chapters 2 and 3 of this book examined, supposed points of interconnection between deep ecology and fascism were typically associated with the development, by some deep ecologists, of the eco-centric orientation of the perspective into an anti-social and misanthropic ideology (Carter 1995; Clark 1996).

Influential in developing an argument for social ecology and in opposition to some of the deep ecological tendencies of primitivists, for example, was Bookchin's *The Ecology of Freedom*. In this text, Bookchin argued that the end of environmental exploitation and destruction could not be effectively achieved without recognition that the problem of hierarchy is at the root of the human domination of both nature and other humans and that neither can therefore be addressed alone. A contrasting position in anarcho-primitivism (also known as anti-civilisationism or anti-civ), as exemplified in the work of the US anarchist John Zerzan, advocates a return to hunter–gatherer-style social organisation, seeing modern tribal societies such as the !Kung and Mbeti people as remnants of a non-hierarchical and non-alienated social utopia (Zerzan 2002). Particularly relevant in the contemporary context of inspiration for modern ecofascists was Zerzan's notorious association with Ted Kaczynski. Known as the 'Unabomber', Kaczynski ran a terroristic bombing campaign against certain university academics in the mid-1990s and is himself recognised as an anti-civ philosopher of sorts. Kaczynski's bombing campaign was ostensibly motivated to prevent the destruction of humanity by killing those whose research he believed would lead to destructive technology such as artificial intelligence. His manifesto, *Industrial Society and its Future,* is a controversial cult text for some eco-anarchists, despite Kaczynski's long-lasting appeal to ecofascism, and his disassociation with and rejection of 'leftism' with respect to its gender and cultural egalitarianism generally and anarchism specifically (Zerzan 2002; Kaczynski 1995).

Another prominent primitivist also influential on contemporary far-right political ecological movements is Derrick Jensen. Like Zerzan, Jensen advocates the destruction of industrial civilisation, but goes further to explicitly defend violence in such a path, especially when employed by oppressed people, likening such actions to those of wild animals protecting their young. Although less committed to the label anarchist or primitivist (Jensen calls himself an 'indigenist'), he has written several important works in the anti-civ canon such as *The Culture of Make-believe* (2002), *Strangely like War: the Global Assault on Forests* (Jensen and Draffan 2003), *Endgame* (2006) and *The Myth of Human Supremacy* (2016).

Debates, then, have been ongoing about the potential influence of deep ecology or anti-civ ideals on contemporary far-right political ideologies often colloquially associated with responses to the destruction of the national environment, such as ecofascism. But it is important to recognise that other streams of anarchist thought were also influential on progressivist environmentalist positions, which themselves provided a bulwark against the reactionary and denialist climate change positions in Australia, which this book has demonstrated to be in certain respects contiguous with an emergent willingness of the far right to accept that climate change is real. For example, the underlying principles of another strain of green anarchism, known as green syndicalism, was foreshadowed in the 'green bans' of the Australian union movement in the 1970s, which supposedly themselves inspired the name of the world's first green party by name, the German Greens (Burgmann and Burgmann 1998). This political orientation seeks to address ecological issues from a working-class anarchist perspective, viewing anarchistically organised unions (such as the Industrial Workers of the World) as fundamental to green anarchist organising, rejecting the notion that environmental concern is a middle-class affectation, and recognising the disproportionate negative effects of environmental harm on lower- and working-class people. Although many anarchist unions have adopted a green syndicalist position, there are not many explicitly green syndicalist major texts, perhaps the sole exception being a work of the founder of the *Radical Criminology* journal, Jeff Shantz, *Green Syndicalism* (2012).

Despite the differences in the ideological persuasion of radical environmental movements in Australia, as globally, radical environmentalism that has genuine ecological concerns at its heart is built around a loose consensus on a variety of things that can be seen as essentially anarchistic. The radical utopian vision of many 'deep greens' is essentially in line with that laid out by thinkers like Kropotkin, Gandhi, Bookchin and all the other anarchists and anarchist-adjacent figures who subscribed to a social vision of small, self-sufficient, directly democratic communities living in harmony with rather than combating their environment. The alternative approach to food production known as 'permaculture', for example, explicitly references Kropotkin's vision of village organisation (Mollison and Holmgren 1978, 4), as well as applying elements of anarchistic philosophy, such as the idea of spontaneous organisation, to the organisation of food production. The anarchistic 'bioregional' vision of an ecological anarchist society that is widespread in contemporary radical environmentalism is not simplistically agrarian, but seeks to rethink cities, industry and other elements of societal organisation in line with Kropotkin's vision set out in the 1899 *Fields, Factories, and Workshops.*

To avoid erroneously conflating some of these ideological tendencies, it is important to first emphasise the crisis conditions to which these radical post-capitalist tendencies respond. Global heating was brought about through hierarchies of exploitation, and its worst impacts will be felt by oppressed peoples. Capitalism is fundamentally rooted in the exploitation of nature and human labour, and this exploitation has led to the ecological crisis we face today (Klein 2015; Moore 2015). These processes also re-enshrine and reinforce global hierarchies related to capital accumulation's reliance upon the appropriation of cheap nature[4] alongside natural resources and human labour. The logic of capitalism to externalise the costs onto others in order to maximise profits has led to the overexploitation of natural resources and the destruction of ecosystems (Patel 2011). The incompatibility of any meaningful measures for global heating mitigation with a capitalist framework, moreover, is exemplified by the latter's contradictory need to prioritise short-term profits over long-term sustainability (Harvey 2006).

While capitalism is not the sole precondition for a rise to prevalence of fascism and far-right social forces, fascist movements have always exploited instability and crisis to gain power. In the current context, not only are the hierarchical social relations combined with perpetual crisis scenarios arising from capitalism at issue, but fascist movements also exploit the droughts, floods and fires in their quest for power (Malm 2020). To be clear, we do not mean to suggest that the solutions to the threat of catastrophic climate change are obvious or straightforward, or indeed that we can discern the ultimately correct path; rather, we assert that it seems highly unlikely that salvation lies in the exacerbation and acceleration of the very things that have created the climate crisis.

It is also necessary to recognise the diverse political applications of post-capitalist anarchistic thinking in global domains of peace and conflict. The NA vision, as exemplified by the autonomous movements in Germany in the US, the UK and Australia (among others), selectively adopted the ideals of more localised and decentralised community organisation along substate lines (Sunshine 2008). Yet the social order affirmed in this vision was antithetical to the core premises of anarchists such as Bookchin, being radically hierarchical in structure and ideological orientation. One example of the political application of Bookchin's work throughout the anarchist world is seen in some of the elements of the Kurdish resistance and political organising during the Syrian Civil War. To briefly elaborate, while imprisoned in Turkey, the leader of the Kurdistan Workers' Party (PKK), Abdullah Öcalan, renounced his Marxist-Leninism and began reading radical literature that might shape a new direction for revolutionary politics in the post-socialist era of the region. Upon discovering Bookchin, Öcalan devoured his entire oeuvre and drew heavily on it to develop a renewed revolutionary vision for the Middle East, called 'democratic confederalism' (Stanchev 2016; Öcalan 2017). Paradoxically, this anarchistic revolutionary vision was imposed on the rank and file of the PKK by its leader; despite this, and although he was imprisoned at the time, the PKK accepted Öcalan's ideological change, which has since been the basis of an ongoing revolution in a Kurdish region of Syria named Rojava. This feminist, ecological, nonreligious, directly democratic federation of autonomous cantons has been for some years the most consistent force opposing the Islamic State group in the brutal civil war in the region. Remarkably, considering his mixed reputation among US anarchists (especially for his reputation as a bitter and noncollegiate critic of other radicals), an entire generation of Kurdish revolutionaries involved in arguably the most significant large-scale application of anarchist principles since the Spanish Civil War cite Bookchin as a direct influence (Knapp et al. 2016).

Trajectories of the environmental far right and fascist connections

In the investigation of this book, we have sought to avoid the pitfall of uncritically accepting the stated ecological and other environmental priorities of *far-right* politically violent actors, or of 'taking them at their word'. Too often in contemporary reporting on the resurgent phenomenon of ecofascism and its political–economic correlates, academics, journalists and policy-makers have also failed to recognise the ideological frameworks of developmentalism that violent actors such as Brenton Tarrant draw upon to justify their actions.

Contextualising both the online and offline political environment in which Brenton Tarrant emerged, the discussion in Chapter 3 showed that his pseudo-environmentalist views were not echoed in the political messaging of even Australia's national socialist leaders such as Thomas Sewell, who approached Tarrant for recruitment. For members of NSN and other extreme-right groups in Australia, climate change was only a marginal topic of discussion, occasionally framed in non-denialist terms; environmentalist attitudes were more consistently directed towards martial combat training outdoors, *Wandervögel* hiking activities and natural landscapes portrayed as a Fatherland coding natural law (see Chapter 3 p. 102). This stood in contrast to the disorganised far-right actors on Gab and Stormfront, who primarily endorsed climate change denialism, with a limited number of exceptions. By their involvement in Australian 'culture wars' rhetoric, the climate change denialism and conspiracy thinking of these far-right actors exerted a 'radial' influence, representing an ideologically salient entry point for vulnerable actors into a more extreme-right neo-fascist core (see Eatwell 1996). Furthermore, given that the glorification of natural environmental settings was important for neo-Nazis and other extreme-right actors in Australia, we suggest that it is appropriate to view the environmental statements of Breivik, Tarrant, Crusius and other homicidal actors as 'canaries in a coal mine' – their violent response to the adverse impacts of climate change potentially serves as a useful early indicator of what could become a more prominent social condition, whether manifest in the actions of loan attackers or in violent ethnonationalist policies enacted by governments.

The analysis presented throughout the chapters also demonstrated how the various elements of crisis response that are vulnerable to co-optation are frequently overlooked in contemporary discussions of far-right climate change politics. Although not all exclusionary and oppressive responses to the effects of global heating may be characteristically referred to as either proto-fascist or directly far-right, it is important to recognise that right-wing social movements do not need to have an exterminationist project to be fascist or to have fascist characteristics (Stanley 2020). This book set out an argument against viewing disparately aligned individual actors and movements in isolation. Rather, it makes the case that in order to properly understand a multidimensional *trajectory* of contemporary right-wing environmental response, it is necessary to appreciate how different strands on this political spectrum relate to and reinforce one another.

Although very few of the political forces and tendencies examined in this book might holistically be described as fascist or even quasi-fascist in nature, when taken together, these disparate positions shed light on the possible advancement of a macro-scale social movement that is rightward-trending and emergent in response to the inevitable devastation brought about by global heating. In this light, it is important to note that the various typologies of political response to climate change examined in this book cover different ideological bases that theorists have described as characterising ideological fascism. For example, the examples explored in our account collectively cover all 13 criteria of Umberto Eco's 'Eternal Fascism' or Ur-fascism, including 'the cult of tradition'; 'syncretism'; 'a rejection of modernism'; 'irrationalism'; the absence of a 'critical spirit'; 'fear of difference'; 'appeal to a frustrated middle class'; conspiracy or 'the obsession with a plot'; which involves portrayal of the enemy as both 'too strong and too weak'; the embrace of an eschatological 'final solution'; 'popular elitism'; 'heroism, selective populism'; and 'Newspeak', referring to the discursive tendencies foregrounded in this book that employ an 'impoverished vocabulary' and 'limit the instruments for complex and critical reasoning' (Eco 1995).

As this book also explored, disparate forms of climate response driven by economic incentives tend to cohere in line with those incentives. Echoing the post-truth ethic currently predominating within the political arenas of neoliberal societies, the different patterns of response tend to be pragmatic and transient in line with achieving instrumentally advantageous outcomes. As Chapter 4's analysis elaborated, New Right lobbying groups in Australia since the 1980s have been influential on government policy and media rhetoric on issues from climate change to Indigenous rights, as well as constitutional matters and industrial relations. New Right political messaging has also adapted to reflect dominant popular discourse in respect of climate change such that it might achieve the most advantageous outcomes, including the protection of the fossil fuel industries. Lavoisier's message, for example, has evolved from outright denial of global heating to denial of its anthropogenic causes, denial of its damaging effects, and eventually to an economically driven rationale of avoiding the devastating consequences and effects of what we know to be true about climate change, given that the mitigatory costs would be too high (Kelly 2019, 178).

In the Australian context, various traditions of response, including significant carbon storage and offset schemes, have been oriented towards support of the fossil fuel industries. This has been justified rhetorically by the enduring legacy of the country's extractive colonial history, and practically in light of the current national economic dependence on mining and industrial agriculture. As Chapter 4 explored, since the 1980s, and intensifying today, New Right and reactionary right-wing propaganda has included concerted campaigns to promote support for fossil fuels in Australia. Neoliberal lobbying on the part of the New Right in Australia has made way for the rise of the far right, or what Moore (1995), drawing on right-wing commentator Gerard Henderson, refers to as the 'lunar right'. It effectively moved the political Overton window rightward, apparently rendering far-right social attitudes less extreme. Towards the end of the twentieth century and early in the twenty-first century, the New Right also served as a primary object of critique and opposition for far- and radical-right reactionaries in Australia, given that its programme of small government, economic internationalism and flexible capital accumulation (Harvey 1990) was seen to undermine the economic and social interests of regional populations, and to represent a disregard of nativist ruralism, along with the erosion of traditional institutions such as the nuclear family (Moore 1995, 134).

Tendencies of opposition to both urban centres and to the interconnected powers of international finance are a dominant trope in far- and extreme-right social movements globally and represent an important aspect of cross-fertilisation between Australian far-right movements and their global counterparts. As the chapters in this book have highlighted, conspiracy about a global financial elite informed the antisemitic agitation leading to the inception of nationalistic regimes across Europe in the nineteenth century. Conspiracies about Judeo-Bolshevism later underpinned the genocidal narratives of the German Nazi administration from the 1930s, and twentieth and twenty-first centuries' neo-Nazis and neo-fascists differentially embraced contemporary conspiratorial narratives of Zionist Occupation Government, Cultural Marxism and the Great Replacement theory.

Far-right conspiracy theories about an alleged global liberal elite persist today, even though powerful far-right political and media organisations and figures, such as Fox News, Sky News Australia and 'populist' politicians worldwide, often promote far-right reactionary ideologies. They typically do so to divert attention from material–historical inequalities (Herman and Chomsky 2010; Roy 2014; West and Buschendorf 2015). In this context, it is essential to note that the term 'populism' is increasingly used in political research on extremism and news reporting as a synonym for far-right political organising. However, this terminology can inadvertently advance far-right agendas while undermining or delegitimising democratic social movements (Mudde and Kaltwasser 2017). It labels both left and right groups as 'populist', thus blurring their distinctions, and often portrays far-right xenophobia and prejudice as organically arising from a largely white, working-class base, rather than resulting from cultural media serving a hegemonic discourse (Mondon and Winter 2020). The modern criticism of 'populism' can also indicate elitist attitudes (Mondon and Winter 2019). Furthermore, the emerging trend of influential political actors advocating radical nativism and anti-immigrant xenophobia partly aims to garner support for future fortress policies in response to the mass displacement of people caused by global heating (Bauder 2016).

Also often omitted in far-right narratives about the supposed tyrannical control imposed by organised elites, is that financial powers lie with right-wing neoliberals and plutocratic actors who ordain themselves at the top of natural hierarchies. Techno-elites such as the billionaire investor Peter Thiel promote 'neo-reactionary' politics through their technological empires, supposedly seeking the instauration of a kind of aristocratic, techno-feudal rule associated with what is sometimes called the Dark Enlightenment (Pein 2018; Rushkoff 2017; Land 2022). In respect of far-right social movements gaining steam in Australia and elsewhere, it is important to note that this contradiction lies at the heart of modern fascism itself. While far-right actors in the Global North – particularly those who advocate for strong borders in response to 'climate refugees' – often outwardly oppose neoliberal international economic systems and those that control them, far-right claims of rightful white supremacist dominance are predicated on the long-term sequelae of global colonial and capitalist projects (Richards 2020b). The mediatised networks through which far-right propaganda is currently disseminated are, moreover, owned, curated and manipulated by techno-elites with personal stakes in the promotion of far-right neo-reaction. These figures also support the continuation of environmentally injurious practices, such as the mining of essential minerals, the conspicuous over-consumption of goods and built-in obsolescence in the physical products they produce (see Benjamin 1935; Sontag 1975; Adorno and Horkheimer 1997; Virilio and Bratton 2006; Stiegler 2018).

As with the other trends and issues examined earlier in this chapter, the contemporary reliance of far-right and neo-fascist movements on technologised networks must be recognised for its historical precedents. For example, the Italian National Fascist Party and German Nazi administration both rejected the supposed cultural decadence associated with modernity, but machines, technology and capitalist industrialisation were always a vital part of the fascist project. Fascism was long recognised as needing modern machines, propagandising mass media and capitalist industrialisation to create the conditions for oppressive and totalising rule (Arendt 1951). Both Mussolini and Hitler built extensive highways and poured financial support into automobile manufacturing, recognising the ways in which capitalist industrialism and the promotion of fossil fuel industries, including coal in particular, could ideologically and technically sustain the machinery of fascism (Tooze 2006; Snyder 2016).

Finally, important in the context of these debates is that while fascism relies upon machines, it is also aptly understood *as* a machine by virtue of its inhumanity, routinisation and mechanisation. It has been described as a form of 'mass psychosis' arising from the alienation and dehumanisation of individuals in capitalist societies, reinforced by psychic machines of mass propaganda (Berardi 2015). It is also understood as a 'war machine' that seeks to destroy all forms of difference by imposing a homogeneous and totalising system of control (Deleuze and Guattari 2009). As Adorno's *Minima Moralia* highlights, moreover, fascism represents the ultimate form of dehumanisation, reducing individuals to mere cogs in a machine-like system of oppression and control. Fascist societies, then, are characterised by a kind of 'mechanical' rationality, in which human beings are stripped of their autonomy and reduced to mere objects to be manipulated and exploited by those in power (Adorno 2005).

The examples provided in this book demonstrate how the mechanistic and reductionistic worldview applied to fascism through ecofascism is also characterised by a form of technological determinism in which the use of advanced technologies and machines, such as those deployed historically in support of fossil fascism, and now evident in neo-colonial and neoliberal greening initiatives, can be seen as the key to achieving human domination over nature. Ecofascism itself represents the aspiration for a totalising and homogeneous system of control over the natural world and human society alike (Merchant 1980). As Biehl and Staudenmaier (1995) argue, ecofascists tend to view the world as a kind of 'resource bank' to be exploited for the benefit of a privileged few, seeking to use technology and industrialisation to dominate and subjugate both nature and human beings.

'Ecological fascism', rooted in a Social Darwinist perspective on humans and the environment, inherently fosters an unsustainable ecological vision, as it is based on destruction and death through oppression and subordination rather than promoting a lasting communal relationship between humans and ecosystems. Critics of industrial capitalism emphasise the incompatibility of perpetual 'growth' with healing the planet or achieving equilibrium. The contradictions of capitalism then contribute to the broader manifestation of fascist ideology, which we follow a Gramscian tradition in viewing as the rawest and most unrestrained form of capitalism. Fascism and the far right create and uphold 'natural hierarchies' that both stem from and are supported by the widespread naturalisation of a rigid ideological belief in hierarchy and competition as essential components of the human condition.

To address the rise of ecofascism and other far-right political environmental ideologies, it is necessary to deconstruct the political and economic environments from which these ideologies emerge. This book has attempted to show that the Australian context, with its colonial legacy of resource extractivism and white supremacy, serves as a prime example of the importance of this context, influencing contemporary engagements with a restricted, exclusionary politics of climate change that has global consequences. However, we have also aimed to highlight some reasons for optimism and propose some potential avenues for change. Australia gave birth to one of the world's first environmental political parties, also inspiring the world's first Green party by name, and numerous grassroots movements and campaigns have been collaborating across communities for years to protect and support the natural environment. Finally, it is important to acknowledge that, regardless of the actions of colonial powers originating from the Global North, Indigenous communities around the world have persevered and will continue their centuries-long struggle to safeguard their homes and territories. Recognising these efforts and looking ahead, we affirm that the challenges posed by climate devastation could be addressed by supporting such localised responses on a larger scale, with a globally collective and inclusive politics of land, place and public connection offering an alternative to far-right segregationism.

Notes

1 Anning was elected under the ONP ticket but sat as a member of Bob Katter's Australia Party and as an independent in the senate from 2017 to 2019.
2 Operation Sovereign Borders (OSB) is a policy introduced by the Australian Government in September 2013, with the primary goal of curbing maritime human smuggling and stopping unauthorised boat arrivals to Australia. It predominantly affects individuals from underprivileged backgrounds who do not have the financial means to access safer, regulated migration routes. Furthermore, OSB perpetuates a system where economically disadvantaged people are increasingly exploited and marginalised by those responsible for transporting them to Australia. This situation is exacerbated by Australia's practice of detaining and processing asylum seekers in offshore facilities.
3 This work predated the often-cited 'first work' of modern environmentalism, Rachel Carson's *Silent Spring*, which covered issues similar to *Our Synthetic Environment.*
4 'Cheap' means that nature is treated as a low-cost input for production, without considering the true costs and consequences of such actions on the environment and future generations. This exploitation often includes deforestation, extraction of non-renewable resources, intensive agriculture, and other practices that lead to environmental degradation and depletion of natural resources.

References

Abbas T, Somoano IB, Cook J, Frens I, Klein GR and McNeil-Willson R (2022) *The Buffalo attack: an analysis of the manifesto*, International Centre for Counter-Terrorism, accessed 1 December 2022. https://www.icct.nl/publication/buffalo-attack-analysis-manifesto

ABF TV (n.d.) 'No way: you will not make Australia home – English' [video], *ABF TV*, accessed 1 July 2018. https://www.youtube.com/watch?v=rT12WH4a92w

Adorno T (2005) *Minima moralia: reflections from damaged life*, Verso Books, New York.

Adorno T and Horkheimer M (1997) *Dialectic of enlightenment*, Verso Books, New York.

Angus I (2016) *Facing the Anthropocene: fossil capitalism and the crisis of the earth system*, NYU Press, New York.

Anstey F (1917) *The kingdom of Shylock*, Labor Call Print, Melbourne.

Anstey F (1921) *Money power*, Fraser & Jenkinson, Melbourne.

Antón-Mellón J (2013) 'The idées-force of the European New Right: a new paradigm?', in Godin E, Jenkins B and Mammone A (eds) *Varieties of right-wing extremism in Europe*, Routledge, London.

Antonucci L, Horvath L, Kutiyski Y and Krouwel A (2017) 'The malaise of the squeezed middle: challenging the narrative of the "left behind" Brexiter', *Competition & Change*, 21(3): 211–229.

Arendt H (1951) *The origins of totalitarianism*, Schocken Books, New York.

Baldwin A, Methmann C and Rothe D (2014) 'Securitizing "climate refugees": the futurology of climate-induced migration', *Critical Studies on Security*, 2(2): 121–130.

Bar-On T (2016) *Where have all the fascists gone?*, Routledge, London.

Bataille G (2015) *On Nietzsche*, Suny Press, Albany.

Bauder H (2016) *Migration borders freedom*, CRC Press, Boca Raton.

Benjamin W (1935) *The work of art in the age of mechanical reproduction*, Penguin, London.

Berardi F (2015) *Heroes: mass murder and suicide*, Verso Books, Brooklyn.

Bevensee E (2020) 'How COVID and Syria conspiracy theories introduce fascism to the left: part 2 – The red-brown media spectrum', CARR, accessed 1 June 2022. https://

www.radicalrightanalysis.com/2020/09/18/how-covid-and-syria-conspiracy-theories
-introduce-fascism-to-the-left-part-2-the-red-brown-media-spectrum/

Biehl J (2015) *Ecology or catastrophe: the life of Murray Bookchin*, Oxford University Press, Oxford.

Biehl J and Staudenmaier P (1995) *Ecofascism: lessons from the German experience*, AK Press, Edinburgh.

Biesecker BA (2018) 'Guest editor's introduction: toward an archaeogenealogy of post-truth', *Philosophy & Rhetoric*, 51(4): 329–341.

Bird D (2014) *Nazi dreamtime: Australian enthusiasts for Hitler's Germany*, Anthem Press, London.

Blaustein J, Fitz-Gibbon K, Pino NW and White R (eds) (2020) *The Emerald handbook of crime, justice and sustainable development*, Emerald Group Publishing, Bingley.

Bloom P (2016) *Beyond power and resistance: politics at the radical limits*, Rowman & Littlefield, Washington.

Boochani B (2018) *No friend but the mountains: writing from Manus prison*, Picador Australia, London.

Bookchin M (1962) *Our synthetic environment*, Knopf, New York.

Bookchin M (1971) *Post-scarcity anarchism*, Ramparts Press, Berkeley.

Bookchin M (1980) *Toward an ecological society*, Black Rose Books, Montréal.

Bookchin M (1982) *The ecology of freedom: the emergence and dissolution of hierarchy*, Cheshire Books, Palo Alto.

Bookchin M (1990) *The philosophy of social ecology: essays on dialectical naturalism*, Black Rose Books, Montreal.

Bookchin M (2007) *Social ecology and communalism*, AK Press, Oakland.

Brown K, Mondon A and Winter A (2021) 'The far right, the mainstream and mainstreaming: towards a heuristic framework', *Journal of Political Ideologies*. https://doi.org/10.1080/13569317.2021.1949829.

Burgmann M and Burgmann V (1998) *Green bans, red union: environmental activism and the New South Wales Builders Labourers' Federation*, UNSW Press, Sydney.

Burgmann V (1984) 'Racism, socialism, and the labour movement, 1887/1917', *Labour History*, 47: 39–54.

Busbridge R, Moffitt B and Thorburn J (2020) 'Cultural Marxism: far-right conspiracy theory in Australia's culture wars', *Social Identities*, 26(6): 722–738.

Busby J (2018) 'Taking stock: the field of climate and security', *Current Climate Change Reports*, 4(4): 338–346. https://doi.org/10.1007/s40641-018-0116-z

Camus R (2012) *The great replacement*, RWTS.

Carter A (1995) 'Deep ecology or social ecology?', *Heythrop Journal*, 36(3): 328–350.

Carter E (2013) *The extreme right in Western Europe: success or failure?*, Manchester University Press, Manchester.

Charkawi W, Dunn K and Bliuc AM (2021) 'The influences of social identity and perceptions of injustice on support to violent extremism', *Behavioral Sciences of Terrorism and Political Aggression*, 13(3): 177–196.

Clark J (1996) 'How wide is deep ecology?', *Inquiry*, 39(2): 189–201.

Conifer D (2018) 'Senator Fraser Anning gives controversial maiden speech calling for Muslim immigration ban', *ABC News*, accessed 7 March 2023. https://www.abc.net.au/news/2018-08-14/fraser-anning-maiden-speech-immigration-solution/10120270

Crock M, Ernst C and McCallum, R (2012) 'Where disability and displacement intersect: asylum seekers and refugees with disabilities', *International Journal of Refugee Law*, 24(4): 735–764.

Curthoys A (1999) 'Expulsion, exodus and exile in white Australian historical mythology', *Journal of Australian Studies*, 23(61): 1–19.

Deleuze G and Guattari F (2009) *Anti-Oedipus: capitalism and schizophrenia*, Penguin, London.

Dixon D (2 July 2021) 'It's happening here: the perils of Sky News After Dark', *The Canberra Times*, accessed 16 November 2022. https://www.canberratimes.com.au/story/7013000/its-happening-here-the-perils-of-sky-news-after-dark/

Doherty B and Hurst D (10 March 2015) 'UN accuses Australia of systematically violating torture convention', *The Guardian*, accessed 1 June 2022. https://www.theguardian.com/australia-news/2015/mar/09/un-reports-australias-immigration-detention-breaches-torture-convention

Dyett J and Thomas C (2019) 'Overpopulation discourse: patriarchy, racism, and the specter of ecofascism', *Perspectives on Global Development and Technology*, 18(1–2): 205–224.

Eatwell R (1996) 'On defining the "fascist minimum": the centrality of ideology', *Journal of Political Ideologies*, 1(3): 303–319.

Eco U (1995) 'Ur-fascism', *The New York Review of Books*, 42(11): 12–15.

Ehrlich PR (1968) *The population bomb*, Ballantine Books, New York.

Escobar A (2011) *Encountering development: the making and unmaking of the Third World*, Princeton University Press, Princeton.

Gentile E (2003) *The struggle for modernity: nationalism, futurism, and fascism*, Greenwood Publishing Group, Westport.

Gerard A and Weber L (2019) '"Humanitarian borderwork": identifying tensions between humanitarianism and securitization for government contracted NGOs working with adult and unaccompanied minor asylum seekers in Australia', *Theoretical Criminology*, 23(2): 266–285.

GID_UKIRE (n.d.) (22 April 2018) 'Generation identity: you will not make Europe home! Back to …' [Twitter], accessed 1 July 2018. https://twitter.com/gid_ukire/status/987721967703904256

Giroux H (2020) 'We must overcome our atomization to beat back neoliberal fascism', *Praxis Educativa*, 24(1): 5–16.

Glaser H (2019) *The cultural roots of national socialism*, Routledge, London.

Gordon SA (1984) *Hitler, Germans, and the 'Jewish question'*, Princeton University Press, Princeton.

Gregor AJ (1979) *Young Mussolini and the intellectual origins of fascism*, University of California Press, Berkeley.

Griffin R (2013) *The nature of fascism*, Routledge, London.

Guha R (2014) *Environmentalism: a global history*, Penguin, London.

Habgood-Coote J (2019) 'Stop talking about fake news!', *Inquiry*, 62(9–10): 1033–1065.

Haidt J (2012) *The righteous mind: why good people are divided by politics and religion*, Vintage, New York.

Hanebrink P (2018) *A specter haunting Europe: the myth of Judeo-Bolshevism*, Harvard University Press, Cambridge.

Hanson P and Merritt G (1997) *Pauline Hanson: the truth – On Asian immigration, the Aboriginal question, the gun debate and the future of Australia*, St George Publications, Parkholme.

Harvey D (1990) 'Flexible accumulation through urbanization: reflections on "postmodernism" in the American city', *Perspecta*, 26: 251–272.

Harvey D (2006) *Spaces of global capitalism*, Verso Books, New York.

Heilbroner R (1974) *An inquiry into the human prospect*, WW Norton & Company, New York.

Herman ES and Chomsky N (2010) *Manufacturing consent: the political economy of the mass media*, Random House, New York.

Hickel J, O'Neill DW, Fanning AL and Zoomkawala H (2022) 'National responsibility for ecological breakdown: a fair-shares assessment of resource use, 1970–2017', *The Lancet Planetary Health*, 6(4): e342–e349.

Horvath R (2013) *Putin's preventive counter-revolution: post-Soviet authoritarianism and the spectre of velvet revolution*, Routledge, London.

Howie L and Campbell P (2017) *Crisis and terror in the age of anxiety: 9/11, the global financial crisis and ISIS*, Springer, New York.

Intergovernmental Panel on Climate Change (IPCC) (2022) *Climate Change 2022*: i*mpacts, Adaptation and Vulnerability*, accessed 19 March 2023. https://www.ipcc.ch/report/ar6/wg2/

Intergovernmental Panel on Climate Change (IPCC) (2023) *AR6 synthesis report: climate change 2023*. https://www.ipcc.ch/report/sixth-assessment-report-cycle/

Jackson R (2015) 'The epistemological crisis of counterterrorism', *Critical Studies on Terrorism*, 8(1): 33–54.

Jacobson D (2019) *Old nations, new world: conceptions of world order*, Routledge, London.

Jensen D (2002) *The culture of make believe*, Chelsea Green Publishing, Vermont.

Jensen D (2006) *Endgame, volume 1: the problem of civilization* (Vol. 1), Seven Stories Press, New York.

Jensen D (2016) *The myth of human supremacy*, Seven Stories Press, New York.

Jensen D and Draffan G (2003) *Strangely like war: the global assault on forests*, Chelsea Green Publishing, Vermont.

Kaczynski T (1995) 'Industrial society and its future', *Washington Post*, accessed 7 March 2022. http://www.washingtonpost.com/wp-srv/national/longterm/unabomber/manifesto .text.htm

Kaldor M and Sassen S (eds) (2020) *Cities at war: global insecurity and urban resistance*, Columbia University Press, New York.

Kelly D (2019) *Political troglodytes and economic lunatics: the hard right in Australia*, Black Inc, Melbourne.

Klein N (2015) *This changes everything*: c*apitalism vs. the climate*, Simon and Schuster, New York.

Knapp M, Flach A, Ayboğa E, Abdullah A and Graeber D (2016) *Revolution in Rojava: democratic autonomy and women's liberation in Syrian Kurdistan*, Pluto Press, London.

Kropotkin P (1912) *Fields, factories and workshops*, Thomas Nelson, New York.

Land N (2022) *The dark enlightenment*, Imperium Press, Baldwin City.

Lindahl S (2018) *A critical theory of counterterrorism: ontology, epistemology and normativity*, Routledge, London.

Lindqvist S (2007) *Terra nullius: a journey through no one's land*, Granta, London.

Lukács G (1952) *The destruction of reason*, Verso Books, New York.

Macklin G (2005) 'Co-opting the counter culture: Troy Southgate and the National Revolutionary Faction', *Patterns of Prejudice*, 39(3): 301–326.

Macklin G (2019) 'The Christchurch attacks: livestream terror in the viral video age', *CtC Sentinel*, 12(6): 18–29.

Malm A (2020) *Corona, climate, chronic emergency: war communism in the twenty-first century*, Verso Books, London.

Mann C (January 2018) 'The book that incited a worldwide fear of overpopulation', *Smithsonian Magazine*, accessed 1 June 2022. https://www.smithsonianmag.com/innovation/book -incited-worldwide-fear-overpopulation-180967499/

Mann ME (2021) *The new climate war: the fight to take back our planet*, PublicAffairs, New York.

Martin G (2015) 'Stop the boats! Moral panic in Australia over asylum seekers', *Continuum*, 29(3): 304–322.

Martini A, Ford K and Jackson R (eds) (2020) *Encountering extremism: theoretical issues and local challenges*, Manchester University Press, Manchester.

McAdam J (2013) 'Australia and asylum seekers', *International Journal of Refugee Law*, 25(3): 435–448.

McChesney RW (2016) *Rich media, poor democracy: communication politics in dubious times*, The New Press, New York.

McCulloch J (2004) 'National (in) security politics in Australia: fear and the federal election', *Alternative Law Journal*, 29(2): 87–91.

McDonald M (2013) 'Discourses of climate security', *Political Geography*, 33: 42–51.

McDonald M (2018) 'Climate change and security: towards ecological security?', *International Theory*, 10(2): 153–180.

McIntyre L (2018) *Post-truth*, MIT Press, Cambridge.

McKay F, Thomas S and Warwick Blood R (2011) '"Any one of these boat people could be a terrorist for all we know!" Media representations and public perceptions of 'boat people' arrivals in Australia', *Journalism*, 12(5): 607–626.

Meadows DH, Meadows, DL, Randers, J and Behrens, WW (1972) *The limits to growth: a report for the Club of Rome's project on the predicament of mankind*, Universe Books, New York.

Mehta L, Huff A and Allouche J (2019) 'The new politics and geographies of scarcity', *Geoforum*, 101: 222–230.

Merchant C (1980) *The death of nature: women, ecology, and the scientific revolution*, Harper & Row, New York.

Minkenberg M (2017) *The radical right in Eastern Europe: democracy under siege?* Springer, New York.

Mollison BC and Holmgren D (1978) *Permaculture one: a perennial agricultural system for human settlements*, Transworld Publishers, Melbourne.

Mondon A and Winter A (2019) 'Whiteness, populism and the racialisation of the working class in the United Kingdom and the United States', *Identities*, 26(5): 510–528.

Mondon A and Winter A (2020) *Reactionary democracy: how racism and the populist far right became mainstream*, Verso Books, Brooklyn.

Moore A (1995) *The right road?: a history of right-wing politics in Australia*, Oxford University Press, Oxford.

Moore A (2005) 'Writing about the extreme right in Australia', *Labour History*, 89: 1–15.

Moore J (2015) *Capitalism in the web of life: ecology and the accumulation of capital*, Verso Books, Brooklyn.

Morris B (2015) *Anthropology, ecology, and anarchism: a Brian Morris reader*, PM Press, New York.

Mountz A (2010) *Seeking asylum: human smuggling and bureaucracy at the border*, University of Minnesota Press, Minneapolis.

Mudde C (2017) *The far right in America*, Routledge, London.

Mudde C (2019) *The far right today*, Polity Press, Cambridge.

Mudde C and Kaltwasser CR (2017) *Populism: a very short introduction*, Oxford University Press, Oxford.

Nairn B (1969) 'The 1916–17 Labor Party crisis in New South Wales and the advent of WJ McKell', *Labour History*, 16: 3–13.

Nelson A (2018) *Small is necessary: shared living on a shared planet*, Pluto Press, London.

Nixon R (2011) *Slow violence and the environmentalism of the poor*, Harvard University Press, Cambridge.

Obaidi M, Kunst J, Ozer S and Kimel SY (2022) 'The "Great Replacement" conspiracy: how the perceived ousting of whites can evoke violent extremism and Islamophobia', *Group Processes & Intergroup Relations*, 25(7): 1675–1695.

Öcalan A (2017) *The political thought of Abdullah Öcalan: Kurdistan, woman's revolution and democratic confederalism*, Pluto Press, London.

Ophuls W (1977) *Ecology and the politics of scarcity: prologue to a political theory of the steady state*, WH Freeman and Company, New York.

Patel R (2011) *The value of nothing: how to reshape market society and redefine democracy*, Granta, London.

Paxton RO (2007) *The anatomy of fascism*, Vintage, New York.

Pein C (2018) *Live work work work die: a journey into the savage heart of Silicon Valley*, Metropolitan Books, Melbourne.

Perera S (2009) *Australia and the insular imagination: beaches, borders, boats, and bodies*, Springer, New York.

Platoff AM, and Knowlton SA (31 August–4 September 2015) 'Old flags, new meanings' [conference presentation], *Proceedings of the 26th International Congress of Vexillology*, Sydney, accessed 3 May 2022. https://steven-knowlton.scholar.princeton .edu/publications/old-flags-new-meanings.

PVVpers (22 Apr 2015) 'PVV: no way: You Will Not Make the Netherlands Home' [video], *PVVpers*, accessed 2 August 2018. https://www.youtube.com/watch?v =wgCSw1JKl7A

Ramirez S (2023) 'Great replacement or slow white suicide?', *Philosophy Today*, 67(1): 177–188.

Richards I (2020a) *Neoliberalism and neo-jihadism: propaganda and finance in Al Qaeda and Islamic State*, Manchester University Press, Manchester.

Richards I (2020b) ''Sustainable development', counter-terrorism and the prevention of violent extremism: right-wing nationalism and neo-jihadism in context', in Blaustein J, Fitz-Gibbon K, Pino NW and White R (eds) *The Emerald handbook of crime, justice and sustainable development*, Emerald Publishing Limited, Bingley.

Richards I (2022) 'A philosophical and historical analysis of "Generation Identity": fascism, online media, and the European new right', *Terrorism and Political Violence*, 34(1): 28–47.

Richards I (2023) 'Capturing the environment, security, and development nexus: intergovernmental and NGO programming for violent and hateful extremism during the climate crisis', *Conflict, Security & Development*, 2023), https://doi.org/10.1080 /14678802.2023.2211019

Richards I, Jones C and Brinn G (2022) 'Eco-fascism online: conceptualizing far-right actors' response to climate change on Stormfront', *Studies in Conflict & Terrorism*. https://doi.org/10.1080/1057610X.2022.2156036

Richards I, Rae M, Vergani M and Jones C (2021) 'Political philosophy and Australian far-right media: a critical discourse analysis of The Unshackled and XYZ', *Thesis Eleven*, 163(1): 103–130.

Richards I, Wood MA and Iliadis M (2020) 'Newsmaking criminology in the twenty-first century: an analysis of criminologists' news media engagement in seven countries', *Current Issues in Criminal Justice*, 32(2): 125–145.

Roose JM (2020) *The new demagogues: religion, masculinity and the populist epoch*, Routledge, London.

Ross AR and Bevensee E (2020) *Confronting the rise of eco-fascism means grappling with complex systems*, Centre for Analysis of the Radical Right (CARR), accessed 1 June 2022.   https://www.radicalrightanalysis.com/2020/07/07/carr-research-insight-series -confronting-the-rise-of-eco-fascism-means-grappling-with-complex-systems/

Roy A (2014) *Capitalism: a ghost story*, Haymarket Books, Chicago.

RT Producers (n.d.) 'Mo Ansar and Martin Sellner debates on RT' [video], *RT Producers*, accessed 25 January 2022. https://www.youtube.com/watch?v=4nG5WEE7m4s

Rushkoff D (2017) *Throwing rocks at the Google bus: how growth became the enemy of prosperity*, Penguin, London.

Rydgren J (ed) (2018) *The Oxford handbook of the radical right*, Oxford University Press, Oxford.

Schiller H (1973) *The mind managers*, Beacon Press, Boston.

Shantz J (2012) *Green syndicalism: an alternative red/green vision*, Syracuse University Press, Syracuse.

Shekhovtsov A (2013) 'European far-right music and its enemies', in Wodak R and Richardson JE (eds) *Analysing fascist discourse*, Routledge, London.

Smith E (2020) 'White Australia alone?: the international links of the Australian far right in the Cold War era', in Geary D, Sutton J and Schofield C (eds) *Global white nationalism*, Manchester University Press, Manchester.

Snyder T (2016) *Black earth: the Holocaust as history and warning*, Crown, New York.

Sontag S (1975) 'Fascinating fascism', *The New York Review of Books*, 6(02).

Southern L (2017) 'The Great Replacement' [video], *LaurenSouthernOfficial*. [Due to the sensitivity of this content we do not include a link to the original content].

Sparrow J (2021) 'Interviewing the far right is bad, so why do journalists keep doing it?: 'no platform' from above and below', *Australian Journalism Review*, 43(2): 177–191.

Stampnitzky L (2013) *Disciplining terror: how experts invented 'terrorism'*, Cambridge University Press, Cambridge.

Stanchev P (2016) 'The Kurds, Bookchin, and the need to reinvent revolution', *New Politics*, 15(4): 77.

Stanley J (2020) *How fascism works: the politics of us and them*, Random House Trade Paperbacks, New York.

Stanovich KE, West RF and Toplak ME (2016) *The rationality quotient: toward a test of rational thinking*, MIT Press, Cambridge.

Staudenmaier P (2021) *Ecology contested*, New Compass Press, Porsgrunn.

Stephensen PR (1935) *The foundations of culture in Australia*. https://nativistherald.com.au /2018/12/01/the-foundations-of-culture-in-australia/

Sternhell Z, Sznajder M and Asheri M (1994) *The birth of fascist ideology: from cultural rebellion to political revolution*, Princeton University Press, New Jersey.

Stiegler B (2018) *Automatic society, volume 1: the future of work*, John Wiley & Sons, Hoboken.

Sunshine S (2008) 'Rebranding fascism: national-anarchists', *The Public Eye Magazine*, accessed 29 November 2022. https://files.libcom.org/files/Rebranding%20fascism.pdf

Tarrant B (2019) 'The great replacement: towards a new society – We march ever forwards', [8chan] [Due to the sensitivity of this content, and legal action in New Zealand, we do not include a link to the original document].

The Parliament of the Commonwealth of Australia (1996) *The Parliament of the Commonwealth of Australia House of Representatives Votes and Proceedings*, 42: 711–724.

Thomas C and Gosink E (2021) 'At the intersection of eco-crises, eco-anxiety, and political turbulence: a primer on twenty-first century ecofascism', *Perspectives on Global Development and Technology*, 20(1–2): 30–54.

Tooze A (2006) *The wages of destruction: the making and breaking of the Nazi economy*, Viking Press, New York.

Tully J (3 December 2015) 'Racism in Australia: from 1788 to stopping the boats', *Green Left*, accessed 1 June 2022. https://www.greenleft.org.au/content/racism-australia-1788 -stopping-boats

United Nations High Commissioner for Refugees (UNHCR) (11 October 2021) *United Nations observations on Australia's transfer arrangements with Nauru and Papua New Guinea (2012-present)*. https://www.unhcr.org/en-au/publications/legal/6163e2984/ united-nations-observations-on-australias-transfer-arrangements-with-nauru.html

Veracini L (2007) 'Historylessness: Australia as a settler colonial collective', *Postcolonial Studies*, 10(3): 271–285.

Virilio P and Bratton BH (2006) *Speed and politics*, MIT Press, Cambridge.

Walker DR and Sobocinska A (eds) (2012) *Australia's Asia: from yellow peril to Asian century*, Apollo Books, New York.

Walker RF (2021) *The emergence of 'extremism': exposing the violent discourse and language of 'radicalisation'*, Bloomsbury Publishing, London.

Ward R (1978) 'Australian legend re-visited', *Australian Historical Studies*, 18(71): 171–190.

Warner J and Boas I (2019) 'Securitization of climate change: how invoking global dangers for instrumental ends can backfire', *Environment and Planning C: politics and Space*, 37(8): 1471–1488.

West C and Buschendorf C (2015) *Black prophetic fire*, Beacon Press, Boston.

Wodak R (2015) *The politics of fear: what right-wing populist discourses mean*, Sage, London.

Wolin R (2019) *The seduction of unreason: the intellectual romance with fascism from Nietzsche to postmodernism*, Princeton University Press, Princeton.

Yakushko O and De Francisco A (2022) 'The (re)emergence of eco-fascism: a history of white-nationalism and xenophobic scapegoating', in Akande A (ed) *Handbook of racism, xenophobia, and populism: all forms of discrimination in the United States and around the globe*, Springer International Publishing, New York.

Zerzan J (2002) *Running on emptiness: the pathology of civilization*, Feral House, Port Townsend.

Zúquete JP (2018) *The identitarians: the movement against globalism and Islam in Europe*, University of Notre Dame Press, Indiana.

Index

For Product Safety Concerns and Information please contact our EU
representative GPSR@taylorandfrancis.com
Taylor & Francis Verlag GmbH, Kaufingerstraße 24, 80331 München, Germany